MASTERING THE MACHINE REVISITED

MASTERING THE MACHINE REVISITED

Poverty, Aid and Technology

IAN SMILLIE

We are becoming the servants in thought, as in action, of the machine we have created to serve us
John Kenneth Galbraith

ITDG PUBLISHING 2000

Published by ITDG Publishing
103–105 Southampton Row, London WC1B 4HL, UK

© ITDG Publishing 1991, 2000

First published in 2000

ISBN 1 85339 514 5 (Hardback)

ISBN 1 85339 507 2 (Paperback)

All rights reserved. No part of this publication may be reprinted or reproduced or utilized in any form or by any electronic, mechanical, or other means, now known or hereafter invented, including photocopying and recording, or in any information storage or retrieval system, without the written permission of the publishers.

A catalogue record for this book is available from the British Library

ITDG Publishing is the publishing arm of the Intermediate Technology Development Group. Our mission is to build the skills and capacity of people in developing countries through the dissemination of information in all forms, enabling them to improve the quality of their lives and that of future generations.

Typeset by J&L Composition Ltd, Filey, North Yorkshire
Printed in Great Britain by Short Run Press Ltd, Exeter

Contents

ACRONYMS ... vii
PREFACE TO THE NEW EDITION ... ix
PREFACE TO THE FIRST EDITION ... xi

PART ONE: THE FAILURE TO LEARN FROM FAILURE

I   A tale of two worlds ... 3
II   Poverty in the South ... 22
III   The best of the West: thinking big ... 35
IV   The third sector and the Third World ... 48

PART TWO: WHAT WE KNOW

V   Technology in history: lies and promises ... 69
VI   Small is beautiful ... 86
VII   Farmers, food and forests ... 104
VIII   Post-harvest technologies ... 121
IX   Energy and power ... 137
X   The house that Jack built: construction materials ... 154
XI   Light engineering and the very late starters ... 171

PART THREE: AN ENABLING ENVIRONMENT

XII   Sustainability: myths and reality ... 189
XIII   Perspectives on women and technology ... 200
XIV   Employment and the informal sector: the economists lose control ... 214
XV   Globalization, adjustment and all that ... 226
XVI   Mastering the machine ... 246

NOTES ... 262
BIBLIOGRAPHY ... 279
INDEX ... 282

Acronyms

ATDA	Appropriate Technology Development Association
ATI	Appropriate Technology International
BRAC	Bangladesh Rural Advancement Committee
BKK	*Bada Kredit Kecamatan*, Indonesia
CIDA	Canadian International Development Agency
DAC	Development Assistance Committee of the OECD
DESCO	*Centro de Estudios y Promoción del Desarrollo*
DFID	Department for International Development
EEC	European Economic Community
EU	European Union
FAO	Food and Agriculture Organization
FCR	fibre concrete roofing
FDI	foreign direct investment
GATT	General Agreement on Tariffs and Trade
GDP	gross domestic product
GNP	gross national product
GRATIS	Ghana Regional Appropriate Technology Industrial Service
GSP	general system of preferences
GTZ	German Agency for Technical Cooperation
HDI	Human Development Index
ICT	information and communication technology
IDRC	International Development Research Centre
IFAD	International Fund for Agricultural Development
IITA	International Institute for Tropical Agriculture
IMF	International Monetary Fund
ITDG	Intermediate Technology Development Group
ITTU	Intermediate Technology Transfer Unit
IUCN	International Union for the Conservation of Nature
KREP	Kenya Rural Enterprise Programme
MCP	mini-cement plant
MEDI	Malawi Entrepreneur Development Institute
MFA	multi-fibre arrangement
MSF	*Médecins sans Frontières*
MYRADA	Mysore Resettlement and Development Agency
NGO	non-government organization
NORAD	Norwegian Agency for International Development
ODA	official development assistance

OECD	Organisation for Economic Cooperation and Development
OPEC	Organization of Petroleum Exporting Countries
OPS	open-pan sulphitation
ORAP	Organisation of Rural Associations for Progress
PRODEM	*Fundacion Para la Promocion y Desarrollo de la Micro-empresa*
SCF	Save the Children Fund
SEWA	Self-Employed Women's Association
SKAT	Swiss Centre for Development Cooperation and Management
TCC	Technology Consultancy Centre
UNCED	United Nations Conference on Environment and Development
UNCTAD	United Nations Conference on Trade and Development
UNDP	United Nations Development Program
UNICEF	United Nations Children's Fund
UNIFEM	United Nations Development Fund for Women
USAID	United States Agency for International Development
VITA	Volunteers in Technical Assistance
VSK	vertical shaft kiln
WWF	Working Women's Forum
WWF	Worldwide Fund for Nature (World Wildlife Fund in the United States)
WTO	World Trade Organization

Preface to the New Edition

ALMOST A DECADE after its appearance, *Mastering the Machine* was still selling well and was still used as a university text in places as far apart as Britain, California and Israel. While much of the original message remained valid, however, many of the things described in the original book had changed dramatically. The rapid growth of communication technologies, for example, has rejuvenated old debates about 'appropriate' technologies, and about how and whether to 'transfer' them to the South. Globalization and the dramatic decline of official development assistance through the 1990s has raised new questions about how the 'developed' world should and will interact with 'developing countries'. And, despite hopeful predictions in World Bank and United Nations publications, world poverty has continued to grow.

Because memories among development organizations are short, the time frame for writing about international development issues is usually confined to a few years, especially where practical, real-life examples are concerned. One of the challenges in preparing a new edition of the book was locating projects that had been started in the 1980s, to see what had become of them. In some cases, projects that looked good had then collapsed. Others have thrived. Both types are described in this edition, giving it a broader time perspective than the first. The book has been substantially re-written, with new material on new technologies (such as communication technologies and photovoltaic cells), and new material on old technologies (such as windmills). There is also a new chapter on sustainability that attempts to clarify its importance and to expose some of the cant that surrounds it.

Mastering the Machine Revisited benefits from a great deal of new source material and from the assistance of many people. I am grateful to Andrew Scott of ITDG in Rugby for all sorts of information, updates and new ideas. Garry Whitby provided updates on Malawi, and Alex Mugova did the same for Zimbabwe, as did Susil Liyanarchchi in Sri Lanka. John Powell, Sosthenes Buatsi, Pamela Branch and Bogdhana Dutka provided new information from Ghana, and Siapha Kamara gathered helpful data on the price of CocaCola in Wa. John Parry in England's Black Country provided updates on his endeavours, and Satish Babu and Paul Calvert assisted with Kerala fishing statistics. Thanks also to Neal Burton of IT Publications, friend, mentor and constructive critic over the past decade and a half. For fifteen years, my son, Rowan, has helped me (frequently) in managing, if not mastering my machines, and more recently in decoding inscrutable attachments from faraway places with strange sounding e-mail addresses. As with the first edition, the opinions contained in the book, as well as any errors or omissions, are mine alone.

Ian Smillie,
Ottawa, February 2000

Preface to the First Edition

For the first time we may have the technical capacity to free mankind from the scourge of hunger. Therefore today we must proclaim a bold objective: that within a decade, no child will go to bed hungry, that no family will fear for its next day's bread, and that no human being's future will be stunted by malnutrition.

Henry Kissinger, World Food
Conference, Rome, 1974

THIS BOOK IS about poverty, aid and technology. It is about a search that has been going on, officially in the developing world for over thirty years, and less officially in most countries since the beginning of time. It is a search that is driven today by more hard-core poverty than has ever before been known, and by a realization that the technology we have applied to solving the problem has severe limitations. We have discovered how to send people safely to the moon, but we have not discovered how to ensure that people can live safely on earth – safe from war, disease, and the crushing poverty that stifles ambition, hope and enterprise.

The book is about the quest for a form of development in the Third World that gives meaning to people's lives, a form of development that is relevant to their resources and needs, and to the hopes people have for their countries and for their children. It is about how what has become known as 'appropriate technology' fits into the larger picture of technology, aid and development, and what it has accomplished in the fight against poverty. The book covers the thinking behind the development of appropriate technology and traces its history through the various changes in development thinking that have characterized the past three decades. Using historical and current examples, it attempts to illustrate the interaction between economic, social and technological factors that give a particular technology a positive impact on poverty reduction. The book is about the possibilities and limitations of technology. It is about failure as well as success, because I believe too many failures in the 'development business' have been ignored or covered up, condemning poor people to suffer the re-invention of too many wheels that never worked in the first place.

It is about large as well as small, for 'appropriate' is not always synonymous with 'small'. It is about policy and people and perversity. It is a study of determination in the face of adversity; and excellence in the face of mediocrity. It is about macro interventions as well as micro interventions, because these are not substitutes for one another, but essential complements to technological progress.

The book is divided into three parts. Part One sets the stage for the following sections by describing the way in which governments and aid agencies, both official and non-governmental, have evolved over the past thirty years in their thinking and approach to issues relating to poverty and development. This section does not deal at length with issues of technology, and may therefore seem somewhat remote from the main theme of the book. It is not. The state of the world and the way its governments and institutions function and relate to one another has everything to do with the technologies they choose to solve their problems. As Fritz Schumacher said, 'the main content of politics is economics, and the main content of economics is technology'. Part Two deals with what has been learned about technology and development, using specific examples and case studies of projects that have succeeded and failed. And Part Three draws lessons derived from many centuries of technological development, but more particularly from the experience of the three 'development decades' that began so hopefully in 1960.

Readers seeking the familiar icons of appropriate technology – biogas digesters, handpumps, latrines and windmills – will perhaps be disappointed that they do not recieve top billing in the pages that follow. It may seem as though Hamlet has been sacrificed at the expense of Rosencrantz and Guildenstern, and that important aspects of appropriate technology have been ignored. The apparent oversight is deliberate. Some 'appropriate technologies', like windmills, represent both old and new solutions to the energy problem. Where they are new, they tend to run a risk observed by Sir Walter Raleigh 400 years ago: 'Whosoever in writing a modern history, shall follow truth too near the heels, it may haply strike out his teeth'.

As for handpumps and latrines, their success in reducing child mortality, raising life expectancy and alleviating the burden on women is well chronicled, and rightly so, by the organizations that have funded their dissemination. Although windmills, handpumps and biogas are here, I have chosen to explore less familiar technological territory in greater detail – cases which may seem less grand and less obvious in the short run, but which are in some ways much more important for the longer-term ability of people to develop and adapt their own technological capacities. To my way of thinking, the Prince of Denmark is alive and well, and lives not in aid agencies, but in the backyard foundries and informal sectors of Asia, Africa and Latin America.

Threading one's way through a business heavily influenced by fads and ideology, it is difficult – especially in a fast moving vehicle – to avoid hitting the occasional sacred cow. In the course of the book a few of these, as well as some white elephants and red herrings, may have been sacrificed. The book also deliberately attempts, perhaps ambitiously, to demystify some ideas that have become lost in a numbing fog of acronyms and jargon. Among them are complex concepts described by once perfectly understandable words: *sustainability, partnership, co-ordination, empowerment, adjustment, transfer of technology, green* and *environment*.

A note on terminology is perhaps warranted at the outset: when the word 'man' is used in this book, it means 'man', not 'men and women'. 'Bilateral donors' refers to the official aid agencies of northern governments, such as the United States Agency for International Development (USAID), or the British Overseas Development Administration (ODA, renamed the Department for International Development, DFID, in 1997). 'Multilateral organizations' are membership institutions comprised of more than one national government. Examples are the World Bank, the International Monetary Fund (IMF) and United Nations agencies such as UNICEF or UNIFEM. The term 'donor agencies' means bilateral or multilateral agencies. For ease of comparison, most values in the book are expressed in United States dollars. Where other currencies are used, such as Zimbabwe dollars, they are clearly marked – Z$.

I have used the word 'billion' in the American sense, meaning one thousand million. References used in only one chapter are fully acknowledged in the notes for that chapter. Those used in more than one chapter are given a brief reference in the chapter and a full reference in the bibliography.

In a sense, the research for the book began twenty-five years ago, when I first went off to teach in a rural school in Sierra Leone. More specifically and more recently, however, my research took me from villages in Gujarat and Bangladesh to potters' sheds in Sri Lanka; from the rolling countryside of southern Botswana to iron foundries in Ghana and roadside workshops in Kenya. It has been a fascinating journey, which included much tea in humble dwellings and equal amounts of coffee in a variety of fortresses: a former Portuguese slaving castle in West Africa, the World Bank in Washington, Oxford University, the University of Sussex, and the United States State Department. I visited workshops in England's Black Country, not far from where the anchor for the Titanic was cast; I talked to teachers in Zimbabwe, tinsmiths in Malawi, and women operating lathes in northern Ghana. I looked at a lot of good and bad handpumps, simple oil presses and clean latrines, and I stumbled through dozens of workshops where some of the world's most genuine entrepreneurs are doing more for their country's economies than battalions of planners and economists.

Many people helped make my work both possible and enjoyable. In Ghana, John Powell, Don Morin, Solomon Adjorlolo, Francis Acquah and Sosthenes Buatsi represent the very many individuals and old friends at TCC, GRATIS and other institutions in Accra, Kumasi, Tema, Tamale and Cape Coast, who took time to get me from place to place and to explain things to me, sometimes more than once. Lynn Wallis provided great assistance and succour with logistics and hospitality. In Zimbabwe, Ebbie Dengu, Jeremy Ascough, Langford Chitsike, David Hancock, Dave Harris, Mike Humphries, Father Brian McGarry, Gibson Mandishona, Thembia Nyoni, Mike Robson, Musa Sithole and Mercy Zinyama all gave me useful perspectives, and most of them also gave me a lot of paper. In Malawi I could not have functioned without the tremendous assistance and hospitality of Garry Whitby, the information and companionship of Keith Machell and the perspectives and photocopier of Don Henry. In Botswana, Tyrell Duncan helped with introductions and insight, and I am grateful to the Botswana Technology Centre and the Rural Industries Innovation Centre at Kanye and others for the time that was taken to discuss and demonstrate their work for me – with special thanks to Kit Morei, Teedzani Woto, Debbie Stayers and Brian Egner. Lahiru Perera drove me up hill and down dale through Sri Lanka's tea estates, while R.M. Amarasekera told me everything I ever wanted to know about fuel-efficient stoves and then sent me on a memorable tour of the brickyards of Negombo. In Kenya, Nick Hall, Nick Evans and Solomon Mwangi took time from busy schedules to talk to me, feed me, and point out things I would not have otherwise noticed.

Without the friendly help, open files and constructive comments on draft chapters by many people at ITDG in Rugby, Nairobi, Dhaka, Colombo and Harare, the book would surely have been impossible. Among them are Simon Burne, Janice Phillips, Frank Almond, Helen Appleton, Barrie Axtell, Alex Bush, Nicky Carter, Adrian Cullis, Peter Fellows, Priyanthi Fernando, Martin Hardingham, Ray Holland, Nurul Islam Khan, David Mather, Patrick Mulvany, Brian O'Riordan, Adam Platt, David Poston, Otto Ruskulis, Theo Schilderman, Andrew Scott, Bob Spencer, Peter Watts, Peter Young and John Young.

I am grateful to Raphael Kaplinsky for an important body of work on appropriate technology, and for putting me onto the 'Third Italy'. Others who helped with time, advice and valuable material include Andrew Barnett, Martin Bell, Dennis Frost, John Howe, Tony Killick, George McRobie, John Parry and his staff in Cradley Heath, and Frances Stewart. Matt Gamser helped in Rugby and Washington, and thanks to frequently crossing paths and the occasional Tusker, David Wright was able to debate both outline and content with me in London, Dhaka, Colombo and Nairobi. He and his family also very kindly put me up (and put up with me) in London. In

Washington I received advice, insight, (and so much paper that an extra suitcase was necessary) from Valeria Budinich, Jean Downing, Carlos Lola and others at ATI, from Beth Rhyne at USAID and from Chris Hennin at the World Bank. Chuck Weiss gave me the benefit of twenty years in the technology business and the people at Development Alternatives Inc. probably gave me a century or two. Parts of Chapters II and XIV are based on earlier work done in collaboration with Roger Young, and parts of Chapter IV are based on research done in conjunction with the Aga Khan Foundation Canada.

I am very grateful to Sharon Capeling-Alakija, Marilyn Carr, David Catmur, Keith Machell, Bill McNeill, Brian Rowe, and Paul Zuckerman for reading various drafts and offering valuable advice. I am also grateful to the very many people in many agencies who helped with information, logistics and advice. Although none of them can take any blame for what follows, they include individuals in the Bangladesh Rural Advancement Committee, CARE, the Canadian International Development Agency, the International Development Research Centre, NOVIB, Oxfam Canada, Sarvodaya, UNDP, USAID and the World Bank. Finally, I would be remiss if I did not say that the opinions contained in the book are mine alone.

Ian Smillie,
Ottawa, July 1991

PART ONE

The failure to learn from failure

The best laid schemes o' mice an' men
Gang aft a-gley
                    Robert Burns

CHAPTER I
A tale of two worlds

Our inventions are wont to be pretty toys, which distract our attention from serious things. They are but improved means to an unimproved end.
                                    Henry David Thoreau, *Walden*, 1854

AT THE BEGINNING of the first millennium, Pliny the Elder wrote in his *Natural History*, 'There is always something new out of Africa'. Today, the reverse is true: there is always something new going *into* Africa. And Asia, and Latin America, and the Caribbean. In 1999, the United Nations Development Program (UNDP) opened three 'technology access community centres' 100 kilometres north-east of Cairo at Zagazig, in Egypt's Governate of Sharkeya. The technology in question was as new as it gets: a server and ten Pentium PCs, plus back-up training computers and dedicated Internet connectivity. UNDP was not alone. By the beginning of the third millennium there were almost 100 donor programmes active in the provision of information and communication technologies (ICTs) in Africa, and there were more in other parts of the world: Acacia, PAN, ChangeNet, HandsNet, SangoNet, SIDSnet. The new pilot project in Egypt was more localized than these. Essentially, it was a 'telecentre' aimed at providing community access to the Internet, and through the Internet to a 'panoply of electronic information and knowledge'. The assumption, or perhaps the hope, was that the telecentre and ICTs more generally would allow faster delivery and better adapted content in long-distance education, telemedicine, environmental management and the creation of new livelihoods. 'ICTs,' UNDP reported, 'can involve more people, hitherto unreached or underserviced ... ICTs allow access to information sources worldwide, promote networking transcending borders, languages and cultures, foster empowerment of communities, women, youth and socially disadvantaged groups ... ICTs are indispensable to realize the global information society and the global knowledge society.'[1]

Maybe. Even probably. But similar things were said about the invention of the printing press, and *exactly* the same things were said about the telegraph in the 1860s. The electric telegraph would 'make muskets into candlesticks'. It 'brings the world together. It joins the sundered hemispheres. It unites distant nations, making them feel that they are members of one great family . . . By such strong ties does it tend to bind the human race in unity, peace and concord . . . it seems as if this sea-nymph, rising out of the waves, was born to be the herald of peace.'[2] In the 1920s the same was said of radio: it would introduce unparalleled advances in education, it would be a powerful force for world peace, it would save democracy.[3]

Among the many lessons about technology in history, perhaps the most important is a warning about hubris: about excessive ambition, pride, arrogance. Among the many lessons about development, the most important is that there are few quick solutions and no easy answers. Among the many lessons about poverty, the most important is that it is greater today than ever before, and it is growing.

From the beginning of time, technology has been a key element in the growth and development of societies. Entire historical eras are named for the levels of their technological sophistication – the Stone Age; the Bronze Age; the Iron Age; the age of sail; the age of steam; the jet age; the computer era. But technology is more than jets and computers; it is a combination of knowledge, techniques and concepts; it is tools and machines, farms and factories. It is organization, processes and people. The cultural, historical and organizational context in which technology is developed and applied is the key to its success or failure. Technology is the science and the art of getting things done – through the application of skills and knowledge.

This book is about a hybrid era, one caught somewhere between bronze and computers, between sail and jet engines; one in which quality, however, has become confused with quantity, and means with ends. For the South – the 'Third World' – it is a time of immense technological opportunity and optimism. It is also a period of unimaginable poverty and hopelessness. It is a time of ferment, change and technological growth. And it is unlike any other period in history, for today – in addition to artisans and artists, farmers, machinists and dreamers – the direction of technology is influenced and fashioned by bureaucrats, economists, far away corporate planners, aid agencies and charities. Never before in history have so many non-technical people exerted so much influence on the advancement, retardation, and movement of technology. This book is about the interaction between these people, and between poverty, aid and technology in developing countries.

Deciding where to start such a complex story is a writer's prerogative, but it can also be a nemesis. Does the story begin with Neanderthal man, inventing simple tools a hundred thousand years ago? Or does it more properly begin ten thousand years ago, when *Homo sapiens* began to domesticate animals and raise crops? Perhaps it begins five thousand years ago with the great civilization of Mohenjodaro in what is today Pakistan. Possibly it begins independently in several places at once – China, Egypt, Mexico, Peru, Zimbabwe, India, Greece, Rome. Perhaps it begins more correctly with colonialism, an age-old two-way purveyor of aid and technology. As a colony, in turn, of Rome, the Norse and the Normans, Britain gained and lost. As a colonizer, it gave and took. Did 'aid' begin with the farming advances taken by the Romans to Gaul and Britain? With Ashoka's conquest of South India? Did concern about credit, debt and development begin with S.S. Thorburn's 1896 report on peasant indebtedness in India? In 1925 with Malcolm Darling's *The Punjab Peasant in Prosperity and Debt*? Perhaps the modern 'aid' era begins with Lend Lease in 1941, or the Marshall Plan in 1948, or the Colombo Plan in 1951.

Chapter V will return to Mohenjodaro, China and Rome. The starting point for the book, however, begins more recently, far from Ancient Rome, and far from the factories and backyard workshops of the Third World. It begins arbitrarily in 1968, with an unlikely collaboration between a winner of the Nobel Peace Prize and a former United States Defence Secretary noted for his escalation of the war in Vietnam. In that year, Robert McNamara, newly appointed President of the World Bank, asked former Canadian Prime Minister Lester Pearson to chair a 'grand assize': an international group of individuals with 'stature and experience' would 'meet together, study the consequences of twenty years of development assistance, assess the results, clarify the errors and propose the policies which will work better in the future'. The report of what became known as the Pearson Commission was presented to the World Bank a year later, and was subsequently published under the title *Partners in Development*.

Since then a number of seminal studies on development have captured the attention of the world and the imagination of some of its leaders. *The Limits to Growth*, the first report to the Club of Rome – an informal international think tank – appeared in 1972.[4] E.F. Schumacher's study of economics as if people mattered, *Small is Beautiful*, was published a year later, and its title quickly found its way into the language and the *Oxford Dictionary of Quotations*. In 1977, Robert McNamara suggested a second international study on development issues, and Willy Brandt, former Chancellor of the Federal Republic of Germany, became its Chairman. Like the Pearson Commission, the Brandt Commission was made up of individuals 'of stature and experience'; unlike the Pearson Commission, it was independent of the World Bank. Although half its funding was provided by the Dutch Government, contributions came from dozens of countries in both the North and the South. In fact it was the title of the Brandt Commission's Report, *North–South; A Programme for Survival*, that gave new expression to the polarity between the industrialized world and what the Pearson Report had quaintly termed 'emerging nations'.

Then came the Palme Commission on security and disarmament issues, followed by the World Commission on Environment and Development, published in 1987 under the title *Our Common Future*, and chaired by the Prime Minister of Norway, Gro Harlem Brundtland.

In recent years, thick reports on the state of just about everything, accompanied by voluminous statistical annexes, have become increasingly fashionable. Each year since 1978, the World Bank has produced a *World Development Report* dealing with major international themes and containing a lengthy series of 'World Development Indicators' – tables which spell out, country by country, the state of the world's economy. UNICEF publishes an annual *State of the World's Children* and the Worldwatch Institute has since 1984 published an annual *State of the World Report* detailing 'progress towards a sustainable society'. In the mid 1980s the World Resources Institute began publishing a series of studies on the management of natural resources and protection of the environment. Late off the starting block but by no means least, UNDP published its first annual *Human Development Report* in 1990 on 'the human dimension of development', a counterpoint or supplement, depending on the perspective of the reader, to the World Bank's more economics-oriented effort.

The season of light; the season of darkness

The Pearson Report, maligned in its day by some as a neo-colonialist apology, relegated now to back shelves in libraries, and long-gone from used-book stores, stands up moderately well. When it appeared in 1969, more than half of the countries of the South had been independent for less than a decade, a decade characterized by optimism and expanding economies. The average annual rate of growth in the South actually exceeded that of industrialized countries. Growth in African export earnings during the 1960s, for example, was equal to that of East Asia and almost four times higher than South Asia. Agriculture, which had stagnated for several years, appeared to be on the threshold of a 'green revolution'. School enrolment in the South had tripled in fifteen years, and great strides had been made in reducing deaths due to plague, cholera and smallpox.

As a product of its time, written by men (there were no women on the Commission) who had lived as adults through the depression of the 1930s, the Pearson Report can perhaps be forgiven for viewing poverty largely in terms of employment, for ignoring women, for simplifying the idea of technology transfer and for the slightly hysterical edge on its approach to the issue of population. The report noted somewhat prematurely that 'Malaria has been virtually eliminated by a worldwide campaign under the leadership of the World Health Organization.'

It is, nevertheless, a prescient document on several counts. It warned against a stagnation in official development assistance, and urged a clearer, more efficient, more developmental approach. It recommended the untying of aid and a downplaying of projects for which recurrent costs would not be available later. It worried about the brain drain from the South to the North, and about the cost and effectiveness of the 100 000 technical experts and volunteers working in the South. It noted the positive contributions of voluntary agencies. It urged a strengthening of multilateral agencies, and it invented the now-famous aid target: 'We recommend that each aid-giver increase commitments of official development assistance to the level necessary for net disbursements to reach 0.7 per cent of its gross national product (GNP) by 1975 or shortly thereafter, but in no case later than 1980.'[5] Warning of growing interest payments on a 1968 Third World debt of $47.5 billion (compared with $1300 billion in 1990 and $2091 billion in 1996), the report suggested a greater grant element, and more concessional loans from aid donors.

It warned of other problems ahead. It gave two estimates of world population by the turn of the century: a low of six billion and a high of seven, based on different assumptions about fertility and life expectancy. That the lower estimate has prevailed should not detract from the alarm with which a 1969 world saw a doubling of its population in only thirty years.

The Pearson Report recognized the problems of growing urbanization and malnutrition, but saw solutions largely in terms of increased investment in jobs. Scale was not a factor in the recommendations except by inference: large problems suggested large solutions. The fast-approaching 'green revolution' was problematic only in the sense that it would require increased levels of investment to support it. Given what transpired in the following decade, one of the Report's more ironic recommendations has to do with foreign private investment as a means of funding some of the development requirements of the Third World. There is none of the concern that would follow in the Brandt Report about the role of multinational corporations. And the basic contradiction between such investment and what the report called 'unsuitable technology' is ignored. Direct private investment is enthusiastically encouraged: 'We recommend the removal, in developing countries, of legal and other barriers to the purchase by institutional investors of bonds issued or guaranteed by governments of developing countries.'[6]

One of the Report's longest sections deals with trade. The first United Nations Conference on Trade and Development (UNCTAD) had been held five years earlier and had made it clear that the development of poor countries would require important adjustments in industrialized countries. In 1969, all but 10 per cent of Southern export earnings was derived from primary products, and half of the countries earned more than 50 per cent of their income from a single primary commodity – jute, sisal, rubber, tin, copper, aluminium, sugar, bananas, coffee, tea, cocoa. It was unequivocally clear to the Pearson Commissioners that diversification and trade were key areas of development potential. The naivety and good sense in their recommendations, viewed across a quarter century of futile negotiations, is almost touching. They recommended the removal by industrialized countries of excise and import duties on primary products from developing countries; the financing of buffer stocks to curtail the damaging effects of price fluctuations; the abolition of quotas during the 1970s; and the development of a general, non-reciprocal scheme of preferences for manufactured and semi-manufactured products from developing countries – before the end of 1970. This, the report states in an ironic choice of words, 'implies a willingness on the part of industrialized countries to make *structural adjustments* which will enable them to absorb an increasing range of manufactures and semi-manufactures from developing countries.'[7] The irony of the prescription is that structural adjustment came to be something imposed by international financial institutions on the South, rather than something required of industrialized countries in the North.

The age of wisdom; the age of foolishness

Ten years later, when Willy Brandt handed his Commission's report to the Secretary General of the United Nations in 1979, the world was a harder, less optimistic place. The post-war boom was at last over. Rocked by one oil crisis and gripped by a second, industrialized countries were in the depths of a recession. 'We are aware', the Commissioners wrote, 'that this report is being published at a time when rich countries are deeply worried by the prospects of prolonged 'recession' and the diminishing stability of international relations ... In the short period since our Commission first met, in December 1977, the international situation has gone from bad to worse. It is no exaggeration to say that the future of the world can rarely have seemed so endangered.'[8]

The Brandt Report, much more hard-hitting than the Pearson Report, made depressing reading. It was deeply concerned about peace, an issue beyond the remit of the Pearson Commission. It understood that environmental issues were of primary concern, and that dependence on non-renewable sources of energy had to be curtailed. It recognized the fundamental importance of women to almost every issue of development. (The Brandt Commission included one woman.) It discussed the unacceptability of Third World poverty in a way that the Pearson report had not.

Brandt wasted little ink on the achievements of the previous decade, because apart from a few spectacular instances, such as the final eradication of smallpox, there was little to report of a positive nature. Although there had been progress in some countries – the Netherlands, Denmark, Sweden, Norway – towards the Pearson aid target of 0.7 per cent of GNP, others (the United States, Britain, France) had actually reduced their aid percentages. 'This failure', Brandt noted, 'points to a marked lack of political will ... Governments with poor aid performances often plead budgetary constraints and balance of payments difficulties, but it is clear that these are not insuperable obstacles. When GNP in industrial countries increases by 3 to 4 per cent a year, the allocation of one-fortieth to one-thirtieth of the annual *increase* in GNP to foreign aid would close the gap between 0.3 and 0.7 per cent in only five years.'[9] Official development assistance spending in 1979 represented about 5 per cent of the $450 billion devoted that year to military spending (by 1996 the ratio had increased to 7 per cent).

In the area of international trade, the decade following the Pearson Report had its promising moments. Although the 1973 oil crisis had a devastating impact on most Third World economies, it also underscored the urgent need for a comprehensive approach to the commodity problem. Years of international lobbying and negotiation culminated in a 1976 UNCTAD IV meeting in Nairobi, which discussed an 'Integrated Programme for Commodities'. This plan formalized and detailed many of the recommendations on trade made in the Pearson Report and proposed a 'common fund' to support buffer stocks and other price stabilization measures. By 1979, however, few of the recommendations had been accepted or ratified by industrialized countries. Tariffs and quotas continued to be the order of the day; prices of primary commodities went up and down like umbrellas on a rainy day. The North basically ignored a series of proposals put forward by the South, and countered with few of its own. It was, the Brandt Commission said, a dialogue of the deaf. 'The air is thick with alibis for inaction ... the result is frustration and deadlock.'[10]

The importance of economic diversification and of increased manufactures in developing countries was not an idea the average Southern economist or planner needed to have repeated. Despite the oil crisis and Northern hostility, efforts in this direction during the 1970s had continued. In 1955, manufactures had accounted for only 10 per cent of non-fuel exports from the Third World. By 1965 the figure had risen to 20 per cent and by 1975 it was 40 per cent. Although these exports originated in a small number of countries, the number of successful exporters was increasing. As with Pearson, for Brandt 'adjustment' to this new situation was a process that *industrialized* countries would have to undergo. Adjustment was not expected to be painless, but it would be far from serious. European studies had already shown that imports from the South were not causing unemployment in the North. A study on adjustment in Britain demonstrated that increased importation of Third World manufactures between 1970 and 1977 had displaced fewer than 2 per cent of the 1970 labour force, while increased exports to the South had created as much new employment.[11]

The 1979 conclusion of the 'Tokyo Round' of the General Agreement on Tariffs and Trade (GATT) talks, however, did little to ease the situation for the South. On manufactured goods the proportion facing OECD tariff and non-tariff barriers in 1979 was 30 per cent, compared with only 11 per cent between OECD countries, and a rate of only 4 per cent in 1974.[12] What this meant in terms of Southern job creation can be seen from the scale of tariffs on different degrees of value added. By 1982, for example, the average tariff barrier in 11 major developed countries on iron ore from the South was zero. It rose to 2.4 per cent on unwrought iron, 9.3 per cent on iron and steel plate, and almost 20 per cent on knives and forks. Unsawn timber entered the North free, but the simple act of sawing it into planks drew 2.2 per cent protection, and turning it into a chair added 11.5 per cent to the protection.[13]

The Brandt Report was much clearer than Pearson on the role of technology in development. It raised questions about farming systems that had been over-mechanized and over-fertilized, at the expense of jobs, the ecology, and ultimately, food production. It echoed other studies that demonstrated the poor results of huge irrigation projects. In India, for example, 72 major irrigation projects commissioned between 1951 and 1965, with a combined command area of 13.4 million hectares, had delivered water to only a quarter of that area by 1966.[14] The 'green revolution' had indeed increased food production in some countries, but often to the accompaniment of impoverishment for smaller farmers. And net food production increases had grown more slowly than the population in 58 out of 106 developing countries between 1970 and 1978.

Many countries in the South, taking their cue from Northern economists, equated development with growth, and growth with industrialization. Massive investments were made in industry on the premise that these investments would form important linkages with other sectors of the economy, by supplying other industries and agriculture, or by increasing the demand for their products as inputs. For some, for a while, the strategy worked, but for many it did not. In Ghana, for example, where the experiment was embraced on a grand scale, manufacturing as a percentage of gross domestic product rose from 10 per cent in 1960 to 14 per cent in 1970.

Most enterprises were highly capital intensive, however, and were completely dependant upon imported raw materials, spare parts and foreign technicians for their survival. Capacity utilization was dismal. Prices rose, exports were not realized and production fell. Yet between 1972 and 1977, the industrial sector sponged up between 42 and 55 per cent of the available foreign exchange – 95 per cent for the import of industrial raw material, and the balance for spares and accessories.[15]

The Brandt Commission pointed out that almost all advanced technology originated in industrial countries where it was developed for a different set of economic and production circumstances. Worse, the North accounted for 96 per cent of the world's spending on research and development, while much of the 'transfer' of technology to the South, or the 'choice' of technology, rested with Northern investors – multinational corporations – and with 'experts' involved in aid projects.

E.F. Schumacher, a professional economist who had worked with John Maynard Keynes and who had been a major intellectual influence on British economic planning during and after the Second World War, had begun as early as 1961 to articulate the challenge of technology in terms of the essential problem it was supposed to solve in the South: poverty. He wrote of the industrial estates that could be found in almost every developing country 'where high-grade modern equipment is standing idle because of lack of organization, finance, raw material, supplies, transport, marketing facilities and the like . . . a lot of scarce capital resources – normally paid from scarce foreign exchange – are virtually wasted.'[16] He spoke of the importance of economic evolution rather than unworkable quantum leaps, and of the need for choosing and developing 'middle' or 'intermediate' technologies. He advocated technologies that could be effectively adopted by those people in the Third World left behind by a modern sector serving and providing affluence to a tiny proportion of the population. Support to large-scale industry would simply worsen the growing problems of rural unemployment and under-employment, and would stimulate further growth of cities, many of them already nightmares of poverty and disease.

The Brandt Commission foresaw two major crises. The first was debt. It noted that the international banking system had successfully channelled OPEC surpluses, in the form of loans, to poor countries, much as Pearson – without foreseeing the oil crisis – had suggested. But the Brandt Commission noted ominously that many countries were in arrears and it spoke of the crisis posed by constraints on the growth of export earnings to meet increased debt service commitments. The second crisis foreseen by the Commission had to do with energy, presaging events such as the Gulf War: 'the political dangers arising from the energy situation are underlined by fears . . . that force might in some circumstances be used by powerful countries to ensure the security of future oil supplies ... putting world peace in jeopardy.'[17]

Despite assurances that the worst did not have to happen, the Report culminated on a gloomy note: 'A number of poor countries are threatened with the irreversible destruction of their ecological systems; many more face growing food deficits and possibly mass starvation. In the international community there is the possibility of competitive trade restrictions or devaluations; a collapse of credit with defaults by major debtors, or bank failures; ... an intensified struggle for spheres of interest and influence, leading to military conflicts. The 1980s could witness even greater catastrophes than the 1930s.'[18]

The spring of hope; the winter of despair

The Brandt Commission recommended a major North–South world summit aimed at an intellectual reorientation, at structural change and increased practical co-operation in the interest of enhanced international peace and development. The meeting, which took place at Cancun in Mexico, in October 1981, was attended by an impressive who's who of leaders and thinkers from both North and South. But Cancun was an inconclusive failure, and industrial countries, still gripped by recession, tightened self-serving measures further.

The 1987 World Commission on Environment and Development, commonly known as the Brundtland Report, added high octane ecological fuel to the increasingly explosive mix of poverty, debt and trade imbalance. In addition to the other calamities heaped on the poor of the South, in the 1970s almost 25 million people had been affected each year by drought and 15.4 million suffered from floods. Unchecked pollution, declining energy and mineral resources, deforestation and plunder of the oceans highlighted the symbiosis between ecology and economics. The Brundtland Report saw – in pollution, in dwindling resources, in the droughts, floods and famines – the clear and devastating hand of man, rapaciously exploiting new-found power to alter planetary systems at a rate that outstripped the capacity of science to assess the effects, and outstripping the capacity of fragmented political systems to check the damage.

Awakened by numerous public attacks on its structural adjustment lending, the World Bank devoted its entire 1990 *World Development Report* to the issue of poverty. What the report revealed was bad. Despite a problem with definitions and statistics, it was clear that the number of poor had increased in both real and relative terms since the Pearson Report. By 1990, a billion people were living in absolute poverty – spending most of what they earned on food, and yet still eating less than enough to remain healthy.

During the 1980s, the Pearson aid target, 0.7 per cent of GNP, had proved as elusive as ever. Few of the large donors came close to a half of one per cent. Even the World Bank was beginning to question bilateral aid effectiveness. It noted that in 1987 41 per cent of all aid went to middle and high income countries.[19] Two thirds of all OECD member countries' aid was tied to their own goods and services, thus reducing effectiveness (or increasing costs) by 20 per cent or more. There was continued emphasis on large, capital-intensive projects rather than support to social sectors. And many aid programmes were much more related to commercial or strategic initiatives than to development. The Bank estimated that in 1986, for example, only 8 per cent of the USAID programme could be identified as 'development assistance devoted to low-income countries'.[20]

'Structural adjustment' had become the Rosetta stone of World Bank and International Monetary Fund thinking in the 1980s. A concept that began life as a potentially thorny problem for the North rather than the South, structural adjustment had, by 1990, changed meaning entirely through a combination of factors. First, primary commodity prices began to collapse at the beginning of the 1980s. ('Collapse', of course, means that the prices Northern importers would pay had collapsed.) Between 1980 and 1982, for example, the price of sugar fell 78 per cent, rubber 37 per cent and copper 35 per cent. Many developing countries were consequently forced to live with fewer resources, lowering consumption unless they had access to foreign capital through loans, investment or aid.

Second, the Brandt warnings on debt began to come true in Poland in 1981, and in Mexico the following year, presaging a debt and banking crisis of massive proportions. As global inflation, Western interest rates and foreign exchange fluctuations all increased, developing countries had to adjust yet again. The economic recession of the early 1980s, the debt crisis and commodity collapse led governments and financial institutions to focus increasingly on short-term problems such as debt servicing, inflation and falling foreign exchange receipts. Most developing countries suffered from a combination of declining exports, rising debt-servicing costs, and severe domestic inflation, with a consequent balance of payments crisis. The doctor's prescription, structural adjustment, referred to the macroeconomic policy reforms required to accommodate these worsening economic conditions. IMF and World Bank structural adjustment loans were intended to reduce the impact of the external shocks by promoting more durable economic structures and by building the foundations for more sustained economic growth. Typically, structural adjustment programmes included currency devaluation, reductions in government spending, and trade and exchange liberalization.

Industrialization and diversification into the export of manufactures –

another aspect of adjustment programmes – assumed access to external markets. Trade liberalization and reduction of tariffs and quotas *by industrialized countries* were therefore important elements of Bank, IMF and OECD discussions on adjustment. In fact, however, it was this element of the adjustment process – not the conditions imposed on and accepted by developing countries – that most often met with failure.

In order to judge how official development assistance and adjustment have affected economic growth and poverty reduction in the South, the entire range of interactions between North and South must ultimately be examined. If developing countries are expected to shift the emphasis of their development policies increasingly towards manufacturing, expansion of trade, and liberalization, there must be concomitant adjustments among developed countries. There must be a rational approach to industrialization and questions of technology. This was clearly recognized by the Pearson and Brandt Commissions; it has underscored the existence of UNCTAD since 1964; it is ritually reinforced by the OECD; it was acknowledged at all GATT meetings and has subsequently been endorsed by the World Trade Organization, established in 1995. For every step forward over thirty years, however, there seem to have been five steps backward. Despite reduced tariff and non-tariff barriers to trade and increasing trade liberalization in the 1980s and 1990s, the share of trade in GDP fell in 44 of 93 developing countries between the mid-1980s and the mid-1990s.[21] Under these circumstances, adjustment runs the danger of becoming not a one-time economic housecleaning, but the continuation of a long downward spiral of Southern belt-tightening.

International concern about the social impact of initial adjustment programmes, particularly their impact on vulnerable groups such as women, children and the poor, led to calls for a different approach. 'Adjustment with a Human Face', is what UNICEF called for in a 1987 study of adjustment programmes, demonstrating that economic reform programmes clearly put social services in jeopardy, especially in Africa. The Bank itself, not yet fully awake to the magnitude of the issue, admitted dozily that 'when structural adjustment issues came to the fore, little attention was paid to the effects on the poor. Macroeconomic issues seemed more pressing, and many expected that there would be rapid transition to new growth paths.'[22] By the late 1980s, international financial institutions were attempting to mitigate the negative impact of reforms on the poor through 'social dimensions of adjustment' programmes. Many, however, were grab-bags of diverse, high-profile welfare schemes; drops of oil for squeaking wheels rather than preventive maintenance for the machine.

In 1990, Michael Manley, then Prime Minister of Jamaica, juxtaposed the problems of adjustment and reform with the debt crisis:

Countries like Jamaica today stand at the threshold of economic growth. In the past, many leaders, myself included, placed too much reliance on government involvement in the production of goods and services. Now ... we have developed strong private sectors, have for years attempted to implement IMF stabilization programmes and World Bank structural adjustment programmes, but find our economies burdened by debt ... I [do not] expect Jamaica to be permitted to walk away from its debt obligations. However, common sense and conscience surely demand that the major powers ... take an urgent look at those agencies' inflexible rules, rules that have long since outgrown justification and now add to the strangling burdens of countries like Jamaica.[23]

The epoch of belief; the epoch of incredulity

The roots of the problems do not lie exclusively in the North or with aid donors, as Michael Manley acknowledged. Many Southern governments have been easily beguiled by promises of the quick development fix, the 'high tech' solution, by unrealistic reliance on Western development models and by the potential for graft that accompanies travelling salesmen and investors. In Africa, the importance of small farmers has been ignored at the ultimate expense of agricultural production, while pricing policies designed to maintain political support from the urban élite have accompanied the downward economic spiral. Rural infrastructure has suffered throughout Africa, and in some countries has all but collapsed.

The arms trade, both a cause and effect of political instability, is a major contributor to Southern debt. Arms sales between manufacturing countries, mainly in the North, and developing countries totalled almost $20 billion in 1996, and the combined defence expenditure of all developing countries was $172 billion – almost one third of what the industrialized countries spent, including the US, Russia, Britain, France and Germany.

Weak public institutions, poor managerial and financial control, and lack of political accountability are in part a cause, in part a result of underdevelopment. Between 1960 and 1986 there were 144 successful coups in the South, half of them in Africa. Between 1945 and 1990 there were 100 wars and civil conflicts in the South, and after that, things got progressively worse. The end of the Cold War and the collapse of communism in Eastern Europe released pent-up floods of disaffection, nationalism and ethnic division. Unpopular regimes were cut loose from their patrons' influence and financial support, and there was a surge of civil wars and uprisings.

Between 1980 and 1995, more than half of the world's poorest countries experienced conflict. Of these, thirty saw 10 per cent of their people dislocated, and in ten of them, more than 40 per cent of the population was uprooted.[24] In the eight years after the fall of the Berlin Wall, four million people were killed in violent conflicts.[25] In 2000, there were more than two dozen major armed conflicts around the world and perhaps two dozen smaller flash points. In 1999, there were 20 million displaced people – half of them in Africa – and there were almost 22 million refugees and asylum seekers.[26] In some countries, an entire generation has grown up in the shadow of war.

Insecurity breeds insecurity: even in countries where peace has returned, long-term private investment remains insignificant, institutions are fragile, trust in government is low, social reintegration is weak. Joblessness among young men – in countries where weapons remain readily available – feeds continuing violence, social dislocation, family breakdown and insecurity.

The end of the Cold War ushered in other changes, including a precipitous decline in aid volumes. The $60 billion figure for 1994–5 represented a 10 per cent drop in real terms in a single year, following declines in several years before that. By 1997, the amount was down to $49.8 billion, representing only 0.22 per cent of the collective GNP of the donor countries, its lowest level in 45 years. Between 1992 and 1997 official development assistance (ODA) from G7 countries fell by 30 per cent in real terms. A modest reversal in 1998 looked respectable in terms of the 1997 figure, but the net decline over the decade remained enormous.[27] The decline is even more precipitous when changes in the composition of ODA are taken into consideration. A significant proportion of ODA – an estimated 12 per cent in 1996 – was going toward emergency assistance, rather than long-term development. More than 4 per cent was diverted from ODA in 1994 to the upkeep of refugees residing in Development Assistance Committee (DAC) member countries. And although some of the $5.6 billion in aid to Eastern Europe in 1996 was not officially counted as ODA, some of it was.

Part of the aid agenda has always been political. The percentage of ODA going to low-income countries has remained constant at about 70 per cent of the total, although the least developed countries receive less than half of this amount. Anomalies abound. Israel and Egypt received almost 21.9 per cent of American ODA in 1995–6, and Palau, whose entire population could be seated in the Houston Astrodome, received more American aid than Bangladesh. New Caledonia and French Polynesia, with a combined population of 410 000 people, received almost 10 per cent of French ODA. And countries with good commercial prospects tend to receive a greater proportion of ODA today than they once did. In 1996–7, China received 5.3 per cent of the world's ODA, almost double what it received a decade earlier.

Several reasons were ascribed to the dwindling of aid resources. Some, such as the end of the Cold War and the subsequent rearrangement of geopolitical priorities, are obvious. After the Cold War, for example, nobody needed Sierra Leone any more, so Israel and Russia closed their embassies, aid plummeted, and a vicious civil war was all but ignored by the world community, even when there were half a million refugees and two million displaced people. In Sierra Leone, as in most of the developing world, the much hoped-for 'peace dividend' never paid off.

Other reasons for aid cutbacks are offered so frequently and so carelessly that they are becoming received wisdom without any serious questioning. One posits a serious decline in public support for aid, a claim contradicted in all OECD member countries by public opinion polls and by a constant growth in private contributions to non-governmental organizations.[28] Another has to do with a perceived need for fiscal restraint in industrialized countries with high deficits. Four decades of doubt about the effectiveness of aid are said to have gained the upper hand. And with the rapid growth of private flows to developing countries, there is a feeling that aid may no longer be necessary.

Given the triumphalism of the market, this last point is worthy of further examination. Net private capital flows to developing countries grew dramatically in 1996, for example, up 30 per cent over 1995 at $243.8 billion, a five-fold increase since 1990. The numbers so far outstripped ODA spending, that it soon became fashionable to predict the 'end of aid', with the private route as the alternative.

There was nothing in the statistics, however, to suggest that this was even remotely possible. First, the term 'net private capital flow' is misleading. It includes bonds, equity investment, foreign direct investment and commercial bank lending. Bond issuance was up by more than 60 per cent in 1996, but if Mexican bonds are subtracted, there was actually a net decline. Foreign direct investment (FDI) represented half of the private flows, and almost half of this went to China alone. Most estimates of FDI in China were, in fact, grossly overstated because of a phenomenon known as 'round tripping', in which domestic funds are sent out of the country and then return to take advantage of tax incentives for 'foreign' investors. A World Bank study calculated a 37 per cent overestimation of foreign direct investment in China in 1994 as a result of this round tripping, and a further 12 per cent exaggeration because of the proclivity of foreign investors to overvalue capital equipment, again for tax purposes.[29]

In all, over 72 per cent of private flows in 1996 went to only 12 countries. Over 21 per cent went to China, and almost 12 per cent to Mexico, much of it for loan rescheduling. Of the 12, only two, China and India, were low-income countries. The top 12 'developing countries' included Russia, Turkey and Hungary, countries ineligible at the best (or worst) of times for ODA. Sub-Saharan Africa received less than 5 per cent of all the flows, and most of this went to South Africa. In fact low-income countries, excluding India and China, received less than 3 per cent of the total.

Private capital is fickle, something Mexico has discovered more than once. As the *Washington Post* explained it, 'computer-savvy profit hunters comb the world in search of better interest rates, cheaper labour and a host of other things, and then swiftly punish inefficient markets or economic missteps.'[30] The World Bank regularly extols the benefits of private flows, but warned, just before the Asian economic meltdown, of their dangers as well: foreign inflows 'can also be seriously destabilizing because they respond quickly to short-run financial turbulence. Recent experience suggests that this turbulence can be contagious, spilling over to other countries and even other regions in ways not necessarily commensurate with the change in risk.'[31]

This was something of an understatement. Private flows to Brazil dropped 23 per cent between 1995 and 1996, and in Hungary they fell by 68 per cent. Thailand was the first Asian domino to totter. Massive foreign investment had led to surplus production capacity, corruption, and a financial system choked with bad loans. In a country where most young people do not attend high school, weak investments in the social sector had led to a serious shortage of skilled workers. The result: a sudden loss of private sector confidence in 1997, a plunge in the value of the currency, a Mexico-style economic crisis, and the need for strong monetary intervention by government. In July 1997, *The Economist* said, 'Thailand faces economic catastrophe.'[32] To forestall this, the IMF stepped in with a $17.2 billion bail-out, and the World Bank arrived with a structural adjustment programme. The irony of the $17.2 billion Thailand bail-out was that it was $2.5 billion higher than the combined $14.7 billion in net foreign direct investment that had arrived in the country over the previous seven years.

And yet as late as September 1997, the IMF – seeing few signs of the 'tensions and imbalances' that usually precede downturns – was predicting that the global economy was about to enjoy its best five-year stretch in 25 years.[33] Thailand, however, was only the beginning. Within a year, Asian tigers had become sick pussycats. Indonesia, in 1997 a paragon of 'good policies' and the world's second or third largest aid-recipient for a quarter century, was reduced to economic rubble. Just before it all went wrong, the 1997 *World Development Report* said that 'Indonesia's broad-based growth has had a spectacular effect on poverty reduction. Between 1970 and 1990 the proportion of the population living below the official poverty line declined from 56 to 15 per cent; other indicators of welfare, such as infant mortality, showed similar improvement'[34] The report missed several weaknesses in the foundation, however: massive corruption, cronyism, human rights abuse, fake democratic institutions, ersatz capitalism, low investments in education and health, high levels of spending on weapons. This was not good development, and if the World Bank didn't know it, the Indonesians who took to the streets in 1998, finally ending the Suharto regime, did.

By then, the World Bank had begun to wake up. In 1999, an internal Bank report acknowledged that Indonesia's rapid growth had created a 'halo effect' in the Bank's dealings with Suharto, and that 'issues of poor governance, social stress and a weak financial sector were not addressed.'[35] Too bad. Too bad for the Bank. Too bad for Suharto. Too bad for the estimated 50 million Indonesians who fell into poverty, despite a $42 billion IMF bail-out.

In a single year, capital flight from Asia was estimated at $75 billion, and overall, net private flows from North to South fell by 28 per cent, almost as much as they had grown the year before.[36] Net international bank lending to Asia fell from $62 billion in 1996 to *minus* $3 billion in 1997,[37] and the combined economies of Thailand, Indonesia, the Philippines and Malaysia shrank by an estimated 10 per cent the following year.[38] Over time, the situation has improved. But where poverty is concerned, it may not matter much, because even in good years, private flows go nowhere near the sectors of focus for most ODA, and very little goes to the countries that are of concern to aid agencies. To suggest that this might change soon, or that private capital flows are in any way a substitute for ODA, is little more than wishful thinking.

Depending on their perspective and the points they want to make, futurologists tend to overemphasize some points and neglect others. In the 1960s, for example, those most worried about the 'population explosion' stressed the likelihood of higher rather than lower growth estimates. The prediction of disaster helped reinforce the demand for drastic action. This approach contrasts with dire warnings of British and French demographers who predicted only 40 years earlier that the population of France would shrink to 29 million by 1980, and the population of England and Wales would fall to 17.7 million by the end of the century.[39]

In fact, population growth rates are falling dramatically in many countries, not because the birth rate has declined, but because AIDS death rates have soared. Zimbabwe's life expectancy dropped from 61 years in 1993 to 49 in 2000 and it will undoubtedly go lower. In Botswana the HIV adult infection rate was 25 per cent at the turn of the century, in Namibia it was 20 per cent and in Zambia 19 per cent.[40] Unlike a declining birth rate which sorts out the demographics over a generation or two, the HIV/AIDS plague is creating havoc across all age groups, all income groups, children, adults, the educated and the illiterate. The closest historical parallel to this holocaust is the sixteen-century smallpox epidemic in the Americas, and the fourteenth-century bubonic plague in Europe.

The Club of Rome's first report, *The Limits to Growth*, published in 1972, contained a variety of doomsday scenarios which showed, depending on the model and time-frame chosen, that a conjunction of unchecked resource depletion, increasing pollution, a finite capacity to produce food, and continued population growth would sooner or later – perhaps as soon as 1990, and probably before 2050 – lead to irreversible global collapse. The report gave substance to the warnings of others, such as U Thant, Secretary General of the UN, who said in 1969 that, 'Members of the United Nations have perhaps ten years left in which to subordinate their ancient quarrels and launch a global partnership to curb the arms race, to improve the human environment, to defuse the population explosion, and to supply the required momentum to development efforts. If such a global partnership is not forged within the next decade, then I very much fear that the problems I have mentioned will have reached such staggering proportions that they will be beyond our capacity to control.'[41]

Because some of what was predicted did not happen, a number of healthy babies were thrown out with the bath water. *The Limits to Growth* suggested, for example, that world mercury resources would be depleted by 1983; tin would be gone by 1985, zinc by 1988, and copper and lead by 1991. It is estimated, however, that known reserves, if used at 1994 consumption rates, will give zinc and lead until 2042, tin until 2053, mercury until 2077 and copper until 2056. In fact, known world copper reserves grew by about one third in 1998 with the discovery of the world's largest copper deposits 35 km south of Kabul. In the year *before* the discovery, copper had already lost one third of its value because of a glut on the market.[42] In the 1940s, the world supply of recoverable crude oil was estimated at about 600 billion barrels. Despite consumption since then, 1995 estimates placed the total at 2300 billion barrels.[43]

World food production, although geographically very uneven, has more than defied the worst predictions of the 1960s. World cereal crops, which provide about half the calories in the human diet, increased 88 per cent between 1965 and 1989, and by more than 100 per cent in the South.[44] African cereal production increased by more than 40 per cent between the early 1980s and the early 1990s.[45] The world may have been late in acknowledging the effects of pollution, but an array of international agreements, control efforts and research, much of it initiated in the 1980s, is beginning to rectify previous mistakes. It might appear, therefore, at least superficially, that the more optimistic futurologists had been at least partially right after all. American architect, engineer and technology guru, Buckminster Fuller, for example, had written in 1967 that 'Humanity's mastery of vast, inanimate, inexhaustible energy sources and the accelerated business of doing more with less of sea, air, and space technology has proven Malthus to be wrong. Comprehensive physical and economic success for humanity may now be accomplished in one-fourth of a century.'[46]

Some of these issues will be dealt with in greater detail in the following chapters. But the truth is that nothing like 'comprehensive physical and economic success for humanity' was achieved in the 25 years foreseen by Buckminster Fuller. Much was accomplished in providing better health and education. Overall, life expectancy in the South increased by 16 years, and adult literacy rose by 40 per cent. But by the end of the century, despite optimistic 1990 World Bank predictions that the numbers would fall, there were another 300 million hungry mouths – a total of 1.3 billion people living in absolute poverty.

Even the most optimistic projections for the future remain bleak. UNDP's annual *Human Development Report* always tries to put on a happy face, listing what has been accomplished (as have some of the paragraphs above) and showing how, with just a little more effort, some of the world's most serious ills could be rectified. Sometimes the report is so unrelentingly positive that it is hard to be anything but optimistic about the future. Buried in the statistics of the 1998 report, however, is a chilling paragraph:

> Continuing economic stagnation and decline in many developing countries poses another formidable challenge to human development. No fewer than 100 countries – all developing or in transition – have experienced serious economic decline over the past three decades. As a result, per capita income in these 100 countries is lower than it was 10, 15, 20, even 30 years ago, depriving their economies of the resources for improving human development.

In 1996, the development ministers and heads of aid agencies that comprise the Development Assistance Committee of the OECD agreed that at the turn of the century, the number of people living in absolute poverty and despair would still be growing. They outlined an agreed course of action that would include greater coherence between aid and other policies that affect developing countries. They agreed to improved co-ordination, and they agreed 'to make mutual commitments with our development partners, supported by adequate resources.'[47] Two years later, little had changed. 'The pace of implementation' said the 1998 DAC Report, 'is a matter for concern and action ... It is time for more vigorous action.' Only Britain and Germany said they would reverse the downward trend in their aid programmes, and Germany subsequently changed its mind, cutting ODA further. Overall bilateral ODA fell by $8 billion between 1996 and 1997, in relative terms a 21 per cent reduction.[48]

The Pearson Commission Report began with the words, 'The widening gap between the developed and developing countries has become the central issue of our time.' More than half of the three decades that followed the publication of that report saw unprecedented growth and prosperity in the North, while increasing millions in developing countries moved further into poverty. If, indeed, the widening economic gap between industrialized countries and the South had become the central issue of our time, the fact seems – in the West at least – to have gone largely unnoticed.

CHAPTER II
Poverty in the South

We have always known that heedless self-interest was bad morals; we know now that it is bad economics.

Franklin D. Roosevelt

THE OBJECT OF much study over several decades, poverty continues to exercise a small army of researchers. Study and research notwithstanding, the data base for measuring and analysing poverty remains inadequate and unreliable. The World Bank's 1990 World Development Indicators contained income distribution figures for only six low-income and 12 lower-middle income countries. By 1997 only a handful of countries were not covered, but the figures remained five to ten years out of date, there was no time comparison, and the data, derived from various sources and using different criteria, were not readily comparable.

The World Bank's 1983 report, *Accelerated Development in Sub-Saharan Africa*, showed six different estimates of growth in gross domestic product for six Sahelian countries during the 1970s. Estimates for Niger, for example, ranged from a low of 0.9 per cent, to a high of 4.7 per cent. Discrepancies ranged from 23 per cent to 1300 per cent. In 1982, the Government of India estimated the percentage of its population living below the poverty line at 37 per cent. The World Bank's estimate was 44 per cent, a difference of over 50 million people. By 1989, the divergence between the Government of India and the Bank had risen to 116 million, more than the populations of Britain and France combined.[1] In 1990, the *World Development Report* said that there had been 280 million people living in absolute poverty in East Asia and the Pacific in 1985. A subsequent 1998 Bank study calculated that in 1987 there were 464 million, an increase of 65 per cent in only two years. In South Asia, however, poverty was shown to have declined, while in Africa it had apparently remained static.[2] Such variations and revisions inspire little confidence in the reliability of statistics.

Part of the problem is an understandable difficulty in gathering accurate data on 'the poor'. The Overseas Development Institute (ODI) has listed nine different 'fault lines' in the definitional debate about poverty.[3] A further problem is what Robert Chambers calls 'the misfit between methods and practice of research.' Many academics and evaluators tend to shy away from poorer regions and poorer people; they avoid the difficult seasons, thus missing important seasonality factors. 'An exaggerated impression of well-being can also be given by the averages beloved of researchers.'[4]

Financial, political, administrative and statistical biases contained in many official surveys may further obscure the bottom end of the poverty picture. Anthropologist Polly Hill observes that statistic gatherers are not necessarily professional, knowledgeable people. Many are under-trained, lonely, under-supervised, over-worked, ignorant, bored, under-paid and strangers to the area in which they work. Their data are 'extracted from unwilling informants by resorting to many convolutions, blandishments and deceits, including sheer guesswork ... Then, at later stages, the field material is commonly fudged, cooked and manipulated by officials at higher levels, the main purpose being to ensure that the trends will be found satisfactory and convincing by those with still greater authority.'[5]

Another unfortunate factor contributing to the problem is a general lack of interest among big development agencies in the impact of their work on poverty. Of the interest groups within a donor agency, few have anything to gain from an investigation of the poverty impact of aid projects. This lack of interest in poverty extends to Southern governments as well; few exert any pressure for investigation of the income distribution effects of aid. Where understanding is concerned, the result is dismal. A 1998 ODI study of European bilateral aid agencies found that few had an explicit poverty-reduction strategy, and that 'very little is known about the impact of agency interventions on the poor since evaluations have either neglected distributional issues or focussed on output rather than impact.'[6]

Measurements and definitions

Poverty, defined in dictionaries as 'the condition of being without adequate food, money, etc.', is usually regarded as a distinct condition: one is either poor or one is not. Attempts to measure poverty, however, and to define the 'poverty line' – the line that divides the poor from the non-poor, have engaged economists and planners for a century, if not two.

The household is often used as the basis for data collection, although this method has its drawbacks. Households of different sizes are not differentiated. Per capita consumption in a large household will differ from that of a small household; thus a low income household need not necessarily be poor. It may simply be small. Differences in consumption patterns between adults, children and the elderly are similarly ignored. Women often eat last, and least. This sometimes severe gender bias in consumption patterns is also obscured in household data. When derived from household data, per capita income levels present difficulties similar to those of household consumption indicators. In addition, variations in seasonal and annual incomes, especially for rural areas, are difficult to capture in statistics, as is income from informal and sporadic sources. Income in one area of a country may buy a very different level of goods from that in another.

Calorie intake has become a common method of measuring poverty: the 'absolute poor' are sometimes defined as those consuming less than 2250 calories per day. When measured on a household basis, however, this technique has the same drawbacks as other household surveys. A lot of reliable information is required, and the objectivity of calorie-based definitions has been challenged. Calorie requirements will vary with age and occupation, may be generally lower in warmer climates, but are higher for pregnant women, lactating mothers and those suffering from infections and parasites. The FAO has attempted to overcome this weakness by calculating an average 'basic metabolic rate' – the energy expended for the function of human organs when a person is in a state of absolute rest.[7] Indicators of health and nutritional status to determine the incidence of stunting and wasting have proven useful in determining the effects of poverty on children. UNICEF, for example, promotes the 'Under Five Mortality Rate' as an important indicator of the state of a nation's children.

Location and concomitant factors such as climate, soil condition and ethnic or religious make-up will also affect the poverty profile. In the Ivory Coast, for example, poverty in the West Forest is, in absolute terms, a quarter of what it is in the Savannah and the East Forest. In the Savannah, where a large percentage of expenditure is devoted to food, non-food consumption indicators show, erroneously, that people there are better off. The same phenomenon holds true elsewhere, most glaringly in the part of India known as the 'poverty square' – Eastern Uttar Pradesh, Bihar, Madhya Pradesh and the neighbouring parts of other states.

Although the use of per capita GNP figures to compare the relative well-being of countries has been widely criticized, and a range of complementary comparative 'development indicators' created, per capita GNP has remained remarkably resilient as the primary measure of comparison. But GNP figures reveal little about the poor; for example, in a hypothetical country of ten people, in which five receive an income of $100 per annum and five receive $10 000, the average annual income would exceed $5000.

A special distinction can be drawn between the very poor in places such as the Indian subcontinent and Africa, where behaviour and problems are considerably different from those for the further 25 to 35 per cent who fall below the 2250-calorie line.[8] The ultra-poor, for example, devote virtually all increases in income to food; other poor people do not. The ultra-poor are much more prone than others to illness, functional impairment, medical risk and inadequate mental and physical performance. These problems make them less able to work and more prone to employment instability. If they are to be assisted, the ultra-poor cannot be lumped into the general category of 'absolute poverty'.

A major difficulty with most measurements is that they are concerned with the *incidence* of poverty, but provide little information on the flow of people into or out of poverty. In addition, there is more to poverty than levels of income and consumption. As important are questions of power, status, recognition, freedom of choice, degree of participation and self-esteem. Given the complexity and the many dimensions of poverty, however, the tendency to over-simplify is perhaps understandable. According to Chambers, 'Planners and politicians need measures to tell them how well (or badly) they and the economy are doing in their attempts to reduce deprivation . . . This makes it all the easier to ignore complex realities and concentrate on single numbers.'[9]

Writing in 1988, Chambers expanded the discussion by dividing deprivation into five dimensions or conditions: poverty proper (lack of income and assets); physical weakness (under-nutrition, sickness, disability, lack of strength); isolation (ignorance, illiteracy, lack of access to government services, remote location); vulnerability (to contingencies, to becoming poorer); and powerlessness:

> To varying degrees these are tackled by government programmes, but vulnerability and powerlessness are neglected compared with the rest . . . Members of elites find physical weakness, isolation and poverty more acceptable and less threatening aspects of deprivation to tackle. They also appear more measurable than vulnerability and powerlessness which are less tangible, and more social and political. So there is a convergence between elite interests in limiting interventions to physical aspects of deprivation, and professional interests in measurability. Together they conspire to concentrate planning, programmes, policy and debate on physical weakness, isolation and poverty, and to neglect vulnerability and powerlessness. Not surprisingly, normal professional and elite paradigms of deprivation are predominantly physical.[10]

Working with a team in gathering background for the World Bank's 2000 *World Development Report*, Chambers refined his thinking further, interviewing a vast range of people living in poverty, and exploring their concepts of well-being and 'ill-being'. Freedom of choice and action, good social relations, security, enough for a good life and physical well-being were paramount. Violence, corruption and lawlessness were key contributors to ill-being. 'Illness, especially catastrophic illness, stands out as a trigger for the downward slide into poverty.'[11]

In 1990, the United Nations Development Programme (UNDP) took some preliminary steps to correct the information problem in its first Human Development Report. It contained what UNDP called a 'Human Development Index' (HDI) which combined national income statistics with adult literacy and life expectancy to provide a more comprehensive overview of life expectancy, education and living standards.[12] This changed the picture contained in GNP figures alone, which had positioned Switzerland, Japan, Norway, the United States and Sweden in the top five rank, in that order. The new figures placed Japan, Canada, Iceland, Sweden and Switzerland in the top rank. Somewhat surprisingly, several developing countries, including Barbados, Cyprus, Uruguay and South Korea, did as well as European countries like Portugal, the Soviet Union and Hungary.

When adjustments were made to the Human Development Index in 1991 for gender-related issues such as female literacy and mortality, the ratings changed again, in some cases dramatically. Japan fell to seventeenth place, while Finland, Sweden, Denmark, France and Norway moved into the top rank. A further calculation is now made to determine *progress* on the human development indicators over time. The 1998 UNDP *Human Development Report*, for example, showed that in some countries – Indonesia, Egypt, Mali, Botswana – the HDI had improved by more than 125 per cent between 1970 and 1995, while in others, such as Thailand, Malaysia, China and Pakistan it had improved by more than 75 per cent. Obviously, those countries near the bottom might make significant progress in percentage terms and still remain low, but even those with higher starting points, such as Mexico, Colombia and Brazil, achieved increases of 20 per cent or more.

Poverty: a profile

Changes and improvements notwithstanding, the reality of poverty cannot be captured in statistical household surveys or bizarre calculations of 'the energy expended for the function of human organs when a person is in a state of absolute rest'. The poor are real people, with real ideas, fears, needs and real hopes for their children, if not themselves. Many find themselves in poverty because sustained drought or ecological degradation has wiped out their animals or damaged improvements to the land made by generations of family members. Women – abandoned, widowed, divorced – and other people impoverished by events, may have moved onto land that is marginal, or even dangerous in terms of soil, weather and ecological conditions. Isolated households in both rural and urban areas are the last to benefit from the flow of goods and services, and the first to suffer when they are diminished. The poor face each day and each new problem in poor health, with few resources, little education and limited knowledge.

It is estimated that two-thirds of the rural population of Africa and one third of the urban population remain below the level of absolute poverty. Infant mortality rates averaged 170 per thousand in 1997, compared with 96 per thousand for all developing countries and only 7 per thousand for industrialized countries. Child death rates in Africa were three times higher than in Latin America. Seventy-five per cent of Africa's urban population and only 39 per cent of the rural had access to safe water, compared with 89 per cent and 62 per cent respectively for all developing countries. Life expectancy was 20 per cent lower in Africa than for developing countries as a whole.

Asia withstood the economic downturn of the late 1970s and early 1980s better than other regions, and several countries experienced reasonable and even impressive growth rates through most of the 1990s. In East and South East Asia population growth declined and there was a steady increase in per capita incomes, with concomitant decreases in poverty. Even before the 1997–98 economic collapse, there were serious problems, however. Asia contains half the world's population, and 75 per cent of all Asians live in China and India, countries with low per capita incomes and growing landlessness. Despite the good news, an estimated one billion people in South and East Asia live on less than a dollar a day.

Although Latin America is the wealthiest Southern region in terms of per capita GNP, almost a quarter of the population lives in absolute poverty. As in Asia, the broad statistics conceal large national and regional variations. In Colombia and Mexico, for example, less than 15 per cent of the population lives in absolute poverty, while in Brazil and Ecuador it is closer to 30 per cent, and in Peru and Guatemala it is 50 per cent. Even within these statistics there are glaring disparities that have worsened over time. In Brazil, the poorest half of the population received 18 per cent of the national income in 1960, but this had fallen to 11.5 per cent by 1995.[13]

There are serious problems involved in estimating rural poverty. In particular, the rural poor are not a homogeneous group. They include tenants, sharecroppers, landless labourers, artisanal fishermen and other vulnerable groups, among whom there are wide gradations of income. Income levels, however, give little indication of how the income is spent – whether on food, clothing or social obligations such as weddings or funerals. People's priorities and strategies for survival change from culture to culture, depending on their economic proximity to desperation.

The positive correlation between land – or security of land tenure – and income has been well documented. Because it is calculated in different ways, the definition of landlessness affects the estimation of numbers. The term may refer to hired agricultural labourers who do not rent or sharecrop; it is sometimes used to describe those occupying an amount of land too small to sustain a household, and it includes those without significant off-farm assets. It is estimated that 10 per cent of Indian households are *totally* landless, and the number in Bangladesh is probably in the vicinity of 30 per cent and rising. If estimates of those living on 'wasteland' and small homesteads were included in these figures, the number of functionally landless would be significantly higher. Estimates of the total landless population of the South, excluding China, place the total at 13 per cent of all rural households, with a further 60 per cent owning too little land to maintain a subsistence living.[14] China withstood the 1997–8 Asian monetary crisis better than other countries in the region, maintaining an eight per cent economic growth rate through 1998. But the good news rested on shaky foundations: there were an estimated 160 million under- or unemployed people in rural areas, and there was an influx of some 40 million people from the countryside into urban areas. Crime statistics give a flavour of the change: in 1994, Chinese courts dealt with 514 000 criminal cases; by 1996 the figure had grown to 1.6 million.[15]

In virtually every country of the world there are more women than men at the lowest levels of income, and generally, households headed by women are among the poorest in society. 'Households headed by women' is a common but misleading term. What it usually means is that the male who would normally be designated as head of the household in most surveys has died, deserted or migrated in order to find work. Female-headed households are poorer than others because there are fewer income earners (in male-headed households, women often add to family income). And women are often employed in work that yields low economic returns. The number of rural households headed by women, particularly in Africa, the Caribbean and Latin America, is increasing. In countries such as Kenya, Botswana and Zambia, a third to one half of the households may be headed by women at any given time.

Poverty imposes a particularly heavy burden on rural women because of their dual role in the economy – both inside and outside the home. An often overlooked feature of Southern agrarian systems, particularly in Africa and Asia, is the crucial role played by women in agricultural production. In much of Africa, where subsistence farming predominates, most tasks associated with subsistence food production and processing are performed by women.

Not all households headed by women are poor, but among the majority that are, poverty may span several generations, passing from mother to daughter. Women are usually less well-educated, have fewer employment opportunities and receive lower wages than men, even where they do similar types of work. They have less access to land, capital and technology, all of which diminishes the efficiency of their production, both in the home and outside. Despite growing gender-awareness among aid agencies, most development programmes aim at improving the productivity of the formal sector, which is populated mainly by male workers. Women continue to rely for income on traditional methods and tend to stay in the informal sector. Consequently differences in labour productivity between men and women continue to widen. In times of economic adversity, women's participation in both the formal and the informal labour markets increases, although their income levels, compared with those of men, remain low. The majority of refugees, worldwide, are women and children.

The multiplier effect of women's poverty is well known: the children of poorer women, for example, are nutritionally less well off than those of women who earn an adequate income.[16] There is increasing acknowledgement of rural women as primary producers, recognition of the obvious links between a food crisis and the marginalization of women, and of the negative impact of excluding rural women from the benefits of technological advancement. Nevertheless, progress in achieving quantitative and qualitative improvements in the status of women remains slow and inadequate.

In most developing countries, children – especially those in the poorest families – assume a wide range of responsibilities before adulthood. For them, absolute poverty is compounded by an unequal distribution of even the least scarce resources. Because of their greater need, poverty places children – particularly young children – at far greater risk than adults. Basically, children in trouble are the victims of adults. They are unable to protect themselves from abuse, or to change the conditions in which they live. They are weaker and more vulnerable than adults, and their plight is clearly the responsibility of adults. Their rights, their problems, and the solutions to their problems lie entirely in the hands of adults.

Beginning with its 1987 *State of the World's Children* report, UNICEF has provided – in its 'Under Five Mortality Rate' – an indication of development based on the welfare, rather than the economic performance, of a large part of the population. The result is a complex picture with two significant features: first, there have been dramatic improvements in the general condition of life in most developing countries over the past three decades; second, millions of children continue to live and die in appalling conditions. And in many cases, the positive trends of past years are being reversed by worsening economic conditions and the techniques chosen to deal with them. An estimated 11.6 million infants and children die each year,[17] half of them between birth and the age of four. Although death rates are generally declining, birth rates remain high, which means that the number of children is high and growing quickly, with as many as one third of them living in absolute poverty.

There is a direct correlation between child mortality and high fertility.

Where child death is a common feature, parents generally expect to have more children. Thus, poverty and high rates of population growth reinforce one another. Historical trends show that population growth rates decline naturally with development, but for many countries the luxury of time no longer exists. Population growth eats directly into economic improvements, in many cases negating them.

It is possible to project the size of a country's eventual stable population using current trends in birth rates. In Bangladesh, for example, it is calculated that the population will rise from 124 million in 1997 to 346 million before it stabilizes. Kenya's population, growing at 3.8 per cent each year, will grow from 28 million in 1997 to 62 million by 2025, and will stabilize at 113 million later in the century. Although the figures are only hypothetical, the trends are not, and the human catastrophe they imply is genuine. Of 49 low-income countries for which comparative data are available, population growth rates actually *rose* in 23 of them between the periods 1980–90 and 1990–5.[18] More must obviously be done to slow or reverse growth rates, and there are concrete examples of what is possible. Growth rates in India, for example, declined from 2.3 per cent annually in the 1960s and 1970s to 1.9 per cent in the 1980s, rising again slightly to 2 per cent in the 1990s. But Japan and China offer more dramatic examples. Between 1949 and 1956, the Japanese population growth rate dropped from 2.2 per cent to 1 per cent, and Japan today has already reached its hypothetical stable population size. In the 1970s, China saw its economic development threatened by a population growth rate of 2.6 per cent, and within a decade, through programmes of incentives, disincentives, improved health and education, and access for parents to family planning services, the rate dropped by half, reaching 1.1 per cent in the first half of the 1990s.

Rural poverty and environmental degradation are closely linked. Traditionally, the technologies and farming practices of rural people have often protected and sustained fragile local environments. But the pressures of population and changing priorities can make major inroads on tradition. Simply put, poverty contributes to the depletion of natural resources, which in turn exacerbates the suffering of the rural poor.

Environmental degradation is manifested in many ways. The search for fuelwood, overgrazing, inappropriate land use, changes in crops and cropping patterns, and population growth have all put serious additional pressures on the land, pressures which lead to erosion, desertification and a depletion of nutrients in the soil. Localized pressures are reinforced by commercial pressures. Large-scale ranching in Botswana, clearance of the rainforests in West Africa and Brazil, commercial logging in Indonesia, Thailand and the Philippines have all led to serious and, in some cases irreversible changes in rural and national environments. Environmental degradation in the Himalayas is thought by some to be a factor in the devastating and unprecedented floods that have taken place in Bangladesh in recent years.[19]

Some of the causes of, and solutions to environmental problems are economic, social or technological; others are political and institutional. For example, in many countries, security of land tenure is a critical element in the willingness of small landholders to change practices that are generations old, to take risks or to make long-term commitments. The basic challenge for the future is not only to halt further degradation, but to restore and improve the productive capacity of the environment so that small farmers can play a useful part in meeting the food requirements of growing populations.

Urban poverty

The statistics on Third World urbanization and misery are as shocking as they are dramatic. The urban population of the South grew from 286 million in 1950 to 2.7 billion in 1998. Pushed off the land by rural poverty, drawn by the lure of better income, the urban poor live in squalid slums, two-thirds of them as squatters, without clean water, access to sewage facilities, medical attention or schools. Many of the slums in which they live are provided with no services whatsoever, and many are regularly razed in the hope that this will discourage more people from coming. It does not. The urban population of the South expands at double the rate of the overall population. Between the mid-1980s and the turn of the century, another *billion* people, representing two-thirds of the total Southern increase, had to be accommodated in cities. By 2025 the global number of urban dwellers will have risen to 5 billion, most of them in the South.

In the poorest countries as much as half the urban population lives in absolute poverty and the numbers are growing. Lagos, with a population of 10 million in 1995, was expected to grow to 24.6 million by 2015. Jakarta was expected to grow from 8.6 million to 13.9 million, and Mumbai from 15 to 26 million. Karachi, already unmanageable with 9.7 million, will grow to 19.4 million – as many people as live in Laos, Cambodia and Papua New Guinea *combined*. Many of today's squalid Southern slums echo Hobsbawm's comment on British cities of the nineteenth century: 'The city was a volcano, to whose rumblings the rich and powerful listened with fear, and whose eruptions they dreaded.'[20]

The correlation between human misery and economic stagnation is abundantly clear: cities with inadequate infrastructure create bottlenecks for economic expansion. Longer travel time from home to workplace, inadequate or costly transport of goods, uncertain delivery of raw materials, inadequate education, housing, child care and health services, environmental decay, and water and power shortages all reduce productivity. In the mid-1990s it was estimated that one million people in Abidjan had no access to piped water and only 30 per cent were serviced by sewers. In the city's slums, infant mortality rates were five times higher than in richer areas of the city. In Jakarta, air pollution had become so bad that 12.6 per cent of all deaths resulted from respiratory tract infections.[21] Cities like these hold great potential for the eruption not only of political unrest, but of uncontrollable disease. The 1994 outbreak of plague in the Indian city of Surat was contained, but it was an object lesson in the possible, just as AIDS has already become a major cause of urban death throughout Africa and Asia.

The future

The 1990 *World Development Report* said that if economic growth continued at then current rates (5.1 per cent in East Asia and 0.5 per cent in Africa); if aid continued to grow; and if the right polices were adopted (policies that harness market incentives and invest in social and political institutions, infrastructure, technology and basic social services for the poor, including primary health-care and primary education) the level of absolute poverty could fall from 1116 million in 1985 to 816 million by 2000. There *was* growth: during the first half of the 1990s in East Asia it was double what the Bank said was needed, and in Africa it was triple. And dozens of countries accepted the 'right policies' prescribed by donors. But aid did not grow, it declined. And it did not target the poorest countries, or the poorest people, or the greatest needs. Only one quarter of bilateral official development assistance went to the least developed countries, and only 10 per cent went to water, sanitation, education, health and population.[22] The Bank's projections were based essentially on strategies very similar to those recommended by the Pearson and Brandt Commissions (and generally ignored by both Northern and Southern governments), and focused on growth rather than development. They ignored the basic fact that when something grows, it becomes bigger. It is perhaps not surprising, therefore, that while economies grew, poverty also grew, and that just before the 1998 Asian economic 'downturn', there were estimated to be 1.3 billion people living on a dollar a day or less, a 16 per cent increase instead of the potential 26 per cent decrease forecast by the Bank. By the end of the decade the Bank's estimate had changed to 1.5 billion.[23]

If, indeed, economic growth is important to poverty reduction, then the Bank's projection was partially invalidated before the ink was dry, at least in India, where half the world's poor live. During the 1980s, India's growth rate was 5.8 per cent, and between 1990 and 1995 it was 4.6 per cent. But much of it was accomplished at the cost of growing budget deficits and increased borrowing. From $20 billion in 1980, foreign debt rose to $90 billion in 1996, and the debt-service ratio tripled. In fact by the end of the 1990s, debt had become a serious issue throughout the South, not just for borrowers, but for lenders facing the prospects of default. In 1996, World Bank and IMF staff examined the 'debt sustainability' of 38 heavily indebted poor countries. They found that eight were unsustainable and 12 were 'possibly stressed'. The rest were deemed 'sustainable' because the ratio of their debt servicing to exports was 25 per cent or less.[24] For the average wage earner, that level of personal debt servicing would be intolerable. When debt servicing reaches 67 per cent of a country's exports, however – as it did with Guinea-Bissau in 1995 – the condition is more than 'stressed', it is terminal. There is probably no precise word to describe the Zambian situation, where debt servicing in 1995 reached 174 per cent of exports.

In 1996 the IMF and the World Bank announced a debt relief plan. After three years, however, only two small relief packages (for Bolivia and Uganda) had been agreed and the rest was essentially stalled. At the 1999 G7/G8 Summit, a bolder plan was announced, aiming to cut $70 billion from the debt mountain facing the 52 poorest countries. Some of this was to be financed from the sale of IMF gold reserves, although only time would tell whether the full amount – still a minuscule fraction of the problem – would ever be reached, and whether the creditors would delve into already decimated aid budgets to finance the write-off.

Although there was no reason in 1990, when the Bank made its hopeful poverty projections, to think that some Southern commodity prices would not rise, there was no particular reason to think they would not fall again either, just as badly as ever. And they did. Between 1990 and 1994, world commodity prices (in constant dollars) fell for tea, sugar, bananas, groundnuts, cotton, jute and sisal. Cocoa remained unchanged. The only ones that showed improvement were coffee, copra, palm oil and rubber, but all of these (and cocoa) earned significantly less than they had in 1984.[25] The real promise, in the absence of the international stabilization measures recommended by Pearson, Brandt, UNCTAD and others, was one of constant instability and price deterioration. There was, in fact, a strong perception in the North, voiced by the American management guru Peter Drucker, that 'the primary products economy has come 'uncoupled' from the industrial economy . . . primary products are becoming of marginal importance to the economies of the developed world.'[26]

What about the next round of predictions? In 1996 the Development Assistance Committee of the OECD set a target: the proportion of people living in extreme poverty in developing countries should be reduced by at least one half by 2015. The following year it gathered together a number of experts to see whether or not the target was attainable, and two World Bank economists presented a rather gloomy prognosis. They based their projections on the need for growth, debunking the Copenhagen Summit's argument that 20 per cent of Southern government and aid expenditure should go to basic health and basic education. The issue, they said, was not increased spending but improved policy and institutional frameworks: 'There is little systematic evidence of public spending on overall health, or on primary health services being associated with better health outcomes.'[27] This throws virtually all previous World Bank and United Nations prognostications about spending on basic health and education out of the window, and more or less justifies the actual low-aid expenditure on these high-rhetoric items. It also reverts the discussion to the old equation: GDP per capita growth equals poverty reduction. The Bank study did acknowledge certain prerequisites for growth, however: an initial head start with an educated populace, plus an absence of revolutions, coups, assassinations, state export monopolies, extreme domestic unrest and 'socialist economic structure[s]'.

This is not unlike the recipe for how to make a small fortune: first, get a large fortune. Needless to say, if you run the right set of data, you will get the result you want. Even in this study, however, if a selection of 36 poor countries were all to change their policy variables from 'poor' to 'good' (i.e. no more assassinations or socialism), nine would still not be able to halve poverty by 2015. In other words, the poverty prognosis is not very good, certainly not good enough to justify continuing along the same gradualist path that has guided governments and aid agencies for the past four decades.

For a large portion of the world's population the question is not so much one of development, as of rescue. It is not so much a question of reducing poverty, as of attempting to halt its spread. To continue business as usual, combining it with band-aids and short-term emergency measures, is to condemn millions of children to an early grave, and increasing numbers of adults to stunted lives of permanent misery. Schumacher acknowledged that poverty had existed from time immemorial. 'But,' he wrote, 'miserable and destitute villagers in their thousands and urban pavement dwellers in their hundreds of thousands – not in wartime or as an aftermath of war, but in the midst of peace and as a seemingly permanent feature – that is a monstrous and scandalous thing which is altogether abnormal in the history of mankind.'[28]

CHAPTER III
The best of the West: thinking big

Growth, like a rising tide, will float all boats

John F. Kennedy

DESPITE OCCASIONAL DIVERSIONS, conventional economic theory of development since the Second World War has been dominated by the concept of growth. It posits that there is a stage in a country's history during which the required conditions for sustained and fairly rapid growth must be consolidated, following which such growth is more or less assured.[1] Growth rates depend upon the level and efficiency of capital investment in physical infrastructure and the industrial sector. More investment in more productive sectors will create higher growth and will speed the development process. Aid can help to consolidate growth and accelerate the development process by filling gaps, raising investment levels and increasing absorptive capacity. Because technology is such an important part of the growth process, aid can be of invaluable assistance in the transfer of technology from the North to the South.

The sorcerer's apprentice

In the old German tale, an apprentice learns the sorcerer's arts, but cannot control the consequences of the spells he casts. In 1969, the Pearson Commission Report, though still uncritically equating development with growth, admitted uncertainty as to what actually promoted growth. 'The growth process is still mysterious, and no decision-maker can be absolutely certain that any of the effects of any particular change in policy such as devaluation, import liberalization, export subsidy, land reform, or labour-intensive rural programmes will produce precisely the effects planned.'[2]

The 'trickle-down' theory held that general economic growth would trickle down to the masses in the form of jobs and opportunities, along with better distribution of other benefits of growth. The theory remained credible until the mid 1970s, when research began to demonstrate that economic growth by itself was no guarantee of a reduction in poverty. While growth had been accompanied by improved equity and poverty reduction in some countries – Malaysia and Indonesia, for example – in others, such as Brazil and Pakistan, it had not. When it became evident that the trickle-down approach was not reaching the poorest well enough or fast enough, two responses emerged. The first, in the early 1970s, led to greater direct focusing of aid projects and programmes on the poor, through a basic needs approach and integrated rural development projects (IRDPs).

Proponents of this approach held that growth was not a sufficient condition for reducing absolute poverty, and that particular initiatives were required to address the special needs of the poor with respect to education, primary health-care, better nutrition and access to safe water. They also argued that poverty-oriented development initiatives were sound economics. Poor farmers are highly responsive; the poor are a resource, and without an educated and healthy labour force, economic progress is impossible.

The limited success of this approach, however, combined with the oil crises of the 1970s and recession in the early 1980s, led to the emergence of a second response, which was essentially a restatement of aggregate growth theories, with new emphasis placed on structural adjustment. With many Third World countries in a state of economic near-collapse, the question became one of restructuring economies so that self-sustaining development would occur.

The debate between those who advocate 'wealth creation' and those who stress 'redistribution' is characterized in much development writing as though the two ideas were mutually exclusive. Social injustices that lead to extremes in the distribution of wealth should, arguably, be altered or removed; but this is not to imply a halt to the creation of wealth. Similarly, the creation of wealth does not have to mean the concomitant creation of work houses and debtors' prisons. In much current development writing, however, strategies for growth and equity continue to be juxtaposed, as though there were always trade-offs to be made in opting for one over the other. Reviewing the relationship between economic growth and poverty alleviation, a major 1988 World Bank Task Force concluded:

> Long term aggregate growth, while necessary, is not sufficient for the alleviation of absolute poverty at the desired speed. Benefits of national growth do not always accrue to those who are below the poverty line. In many instances, the poor do not possess the physical and human assets which would bring them higher returns ... This helps explain why the number of people living in absolute poverty has increased in low-income countries even while respectable growth rates were achieved over the past three decades ... Areas of complementarity between the twin objectives of efficient economic growth and the alleviation of poverty exceed those areas involving serious trade-offs. Pursuit of both objectives is therefore possible, indeed essential to lasting, stable and humane development.[3]

Although it is widely accepted that growth-oriented strategies must be supplemented by clearly defined poverty-alleviation efforts, by far the greater emphasis in development spending remains on growth.

The 1987 Brundtland Report observed that sustainable development requires more than growth; it requires a change in the *content* of growth, making it less dependent on energy and making it more cognizant of the capital stock on which it depends. The report used the example of income from forestry operations. Only the value of timber extracted and the cost of extracting it are measured in national accounts. The cost of regeneration, or the losses accruing from non-regeneration are ignored. Depreciation, commonly calculated in national accounts, is not extended to minerals, timber, oil, gas and other natural resources. Countries dependent on natural resources are, therefore, in effect, mortgaging the future in order to pay for today's growth statistics. The Brundtland Report concluded that 'in all countries, rich or poor, economic development must take full account in its measurements of growth of the improvement or deterioration in the stock of natural resources.'[4]

The Washington-based World Resources Institute took the process a step further in a calculation of what such an approach might mean in Indonesia. Often cited through the 1980s and 1990s as a developmental success story, between 1971 and 1984, Indonesia's gross domestic product grew at an annual rate of 7.1 per cent, higher than that of most OECD member countries. However, by subtracting estimates of net natural resource depreciation in only three sectors – oil, timber and soils – the study reduced the growth rate to 4 per cent.[5]

If other factors had been taken into consideration, such as the future effect on agricultural potential of soil erosion, the figures would have been reduced even further. In Indonesia, the erosion of hillsides on Java, Bali and Madura averages 54 metric tons per hectare each year, a problem far from exclusive to Indonesia. In the United States in the 1980s, the government converted 13 million hectares of farmland to woodland and grassland before it became wasteland, such was the rate of topsoil loss.[6] In such instances, the cost in lost production is a calculation that is not difficult to make. Similar calculations could be made with regard to lowering of water tables, and for pollution such as acid rain, which eats into the income of future generations.

Infrastructural development

All countries trade. All countries import things they do not produce, and they export commodities and products for which there is an external demand. For the South, especially for the majority of countries without oil, the terms of trade have deteriorated steadily over thirty years, while access to Northern markets has been systematically blocked. Many developing countries concentrated on import-substitution by placing high tariffs on imports in order to encourage the development of local production. But those countries that embarked on aggressive import-substitution policies usually found that extended protection for local production, especially where imported raw materials were required to sustain an industry, led to inefficiencies, inflation, corruption, and in extreme cases to economic collapse.

Northern commercial bank lending for investment all but disappeared with the debt crisis of the 1980s, while structural adjustment lending and IMF stabilization programmes became increasingly conditional on opening economies to more, rather than fewer competitive forces. Developing countries were urged, therefore, to exploit their comparative advantages, instead of investing in risky new areas. As Chapter I showed, however, efforts to exploit the most competitive Third World advantage, cheap labour, resulted for many years in an effective Northern blockade against Southern manufactures, most notably textiles.

Many countries have exploited cheap labour in another way, by taxing agricultural production. Because revenue for public spending is low in countries where the tax base is small, this makes sense at one level. Inevitably, however, high agricultural taxes simply discriminate against the producer, ultimately reducing production. Great emphasis has therefore been placed on internal growth-oriented policies and programmes, and on efforts to increase production for internal consumption, particularly in agriculture. Having failed to achieve anything like a world trading system to match the rhetoric used in GATT, the WTO or UNCTAD, the North – in the form of bilateral and multilateral aid agencies – joined this battle for growth and development in the fields and cities of the South.

Infrastructural development in many parts of today's South – the complex irrigation structures of Syria or Sri Lanka, for example – predates Ancient Greece and Rome by centuries, and has continued up to the present day. It has been an especially intense area of endeavour, however, since the days when aid agencies first arrived in the South. Irrigation, roads, railways, communication facilities, schools, water systems, hospitals, have all been seen as essential precursors to, or components of, growth and development. There are myriad studies to show that, as a generic approach, infrastructural improvement is important to development, and few would dispute the contention. A study of 58 countries over the period 1969–78 found that a 1 per cent increase in irrigation, for example, could be associated with an aggregate increase in crop output of 1.6 per cent.[7] A Bangladesh study showed that households in villages benefiting from public infrastructure – roads, schools, banks, electricity – had significantly improved income from better crop production and more wage labour than villages which had not been exposed to the new services.[8] The oil-related deterioration of the Ghanaian road network during the 1970s had profound effects on the overall economy: journeys took three times longer than necessary, at a significant additional cost in fuel. Bad roads caused more frequent vehicle breakdown and write-off. Shortage of vehicles made distribution of food and agricultural inputs such as seed and fertilizer difficult or impossible. Lack of transportation to move cocoa to the port meant that exports, and therefore foreign exchange declined. Roads were the dog that chased the cat, that ate the rat, that killed the mouse that lived in the house that Jack built.

The levels of investment, efficiency, and technology behind any infrastructural development are critical to its success and, like the measurement of growth, to the all-important question of *who* actually benefits from it. The Bangladesh transport sector is an example of infrastructural development that has consistently avoided dealing with these questions.

At the time of Partition in 1947, Bangladesh – then East Pakistan – had only 287 miles of paved road, mostly in towns and cities. An urgent programme of road and bridge building was undertaken, and by 1970, the network of paved roads had increased to 2000 miles. Damage during the War of Liberation, however, was enormous: 270 road and rail bridges were destroyed or damaged; rail and rolling stock were destroyed, river routes and port facilities were blocked or wrecked. Logically, the transport sector became a focus for the large aid efforts that followed the independence of Bangladesh in 1971.[9]

By 1974, most of the damage had been repaired, but investment in the transport sector continued to grow. Because of the alluvial soil, the vast number of rivers and the annual floods, road building in Bangladesh is enormously expensive – 60 per cent higher, for example, than in India. During the 1970s the transport sector absorbed 17 per cent of public expenditure, and by 1980 it was estimated that on-going and planned road construction would occupy the country's construction industry fully until the end of the century. The highly inefficient railway sector had by 1980 become, in the words of one study, 'a millstone around the government's fragile financial neck'. The World Bank warned strongly against any fleet expansion and recommended 'drastic action'. Nevertheless, road construction, railway fleet expansion, the importation of cars, trucks and buses, and the establishment and growth of an international airline continued apace, much of it funded through general balance of payments support made available by international donor agencies – 30 per cent of it through bilateral aid programmes.[10] Canadian aid to the railways – largely in the form of locomotives and rolling stock – was, through most of the 1980s and 1990s, Canada's largest aid project anywhere.

Considering most Bangladeshis' experience of transportation, the situation was ludicrous. In 1986, nearly 94 per cent of all commercially operated vehicles and boats in Bangladesh were non-motorized. There were four times more rickshaws than all motorized vehicles and boats combined, and for every motorized vessel, there were 100 traditional wooden 'country boats'. These country boats provided 60 per cent of the employment in transport, and the combined carrying capacity of country boats, rickshaws and bullock carts was double that of the mechanized sector. The reason was simple: 70 per cent of Bangladesh's villages were at least five and as much as ten miles from access to any form of motorized transport.

This is not to suggest that roads and railways should be ignored; rather it is to point out the overwhelming indifference of government and aid agencies to the importance of the non-mechanized sector. As one study points out, in their approach to the transport sector, government and aid agencies 'effectively exclude 95 per cent of all vehicles and craft . . . [which] account for 80 per cent of all employment in a country beset by chronic and worsening unemployment, and account for 75 per cent of the value-added by transport operations.'[11] In other words, the transportation needs of the poor in Bangladesh are more or less ignored. Of all projects in transportation, government-funded or otherwise during the 1980s, only 1 out of 334, representing 0.004 per cent of all public investment in the transport sector, had anything to do with non-motorized transport.

The dam busters

Massive dam projects are another form of infrastructure development. With varying emphases on irrigation, flood control and power generation, they have captured the imagination of planners and absorbed great amounts of development spending over the past forty years. Dams offer poor countries a one-off technological solution to some of the most intractable of development problems. Results, however, have been very mixed in most cases, and disastrous in many. Egypt's Aswan High Dam, inaugurated in 1970, had all the usual objectives of a large dam project: power generation, irrigation and flood control. It improved the lives of many poor Egyptians, but the side effects that came with it had not been anticipated: pollution of drinking water because of increased requirements for fertilizer, destruction of the country's sardine industry because of disruption to the aquatic food chain in the Nile Delta, and the spread of bilharzia.

The Akosombo Dam in Ghana, completed in 1972, has rarely functioned to projected capacity and created a wide range of ecological and human problems in the catchment area. Worse, however, were the downstream effects. In the three decades after the dam's commissioning, a great swath of beach in neighbouring Togo disappeared, damaging roads, port facilities, fishing villages and threatening Lake Togo – an important wildlife habitat and source of human drinking water. The Kainji Dam in Nigeria halved fishing and dry-season harvests from traditional floodplain agriculture.[12] Perhaps the most extreme cases of growth-oriented dam projects are those built for Soviet cotton irrigation, reducing the flow of the rivers feeding into Central Asia's Aral Sea. Once the world's fourth largest freshwater lake, the Aral sea has lost more than 75 per cent of its volume since 1960, the equivalent of draining both Lake Ontario and Lake Erie. Today the Aral Sea is a polluted, saline disaster area. Its once vibrant fishing industry has disappeared, and with it, 60 000 jobs.[13]

India's infamous Narmada Project became so controversial that the World Bank eventually withdrew from the project. Although the Indian Government vowed to finish it, it is unlikely that the Sardar Sarovar Dam, which is at the heart of the project, will ever be completed.[14] The massive environmental damage done by the Itaipu Dam on the Parana River between Brazil and Paraguay and by the gigantic Mahaveli scheme in Sri Lanka, funded by Germany, Canada, Britain and the United States, can hardly be calculated. In human terms, 50 000 people in the former were forcibly relocated, many without adequate compensation. In Sri Lanka, the combined resettlement and new settlement projections exceeded a million people, and in the mid-1980s became a destabilizing factor that contributed to the country's devastating civil war. Between 1979 and 1985, 40 World Bank-financed projects for agriculture and hydro-electric power development resulted in the forced resettlement of 600 000 people in 27 countries, with devastating consequences for their health and livelihoods.[15] It should be pointed out that many of these projects, if not most, were conceived and requested by Southern governments, not the development agencies and lending institutions. This does not, however, absolve banks and donors from complicity, especially when the technology, the expertise and the final decision in almost all cases lay in the North, rather than the South.

Perhaps the most controversial dam project is one that has been on the design board longest. The Three Gorges Dam on the Yangtze River promises to be the world's biggest-ever power development. It was proposed by the Chinese Government as far back as the 1920s, but adequate funding and technical capacity did not exist until the late 1980s. An $11 million study, carried out under the auspices of the Canadian International Development Agency between 1986 and 1989, examined the Chinese proposal as part of its aid programme, and recommended that the project go ahead.

A Canadian NGO, Probe International, formed part of a growing lobby against the dam. Forced to resort to Canada's Access to Information Law, Probe eventually obtained a copy of the 13–volume CIDA report, and submitted it to an international panel of 11 eminent professionals and academics in the fields of engineering, economics, environment, hydrology, biology and soil science. Given their environmentalist credentials, the outcome was perhaps predictable. The results were published by Probe International in book form in 1990: *Damming the Three Gorges; What Dam-Builders Don't Want you to Know*.[16] The book investigated the Canadian findings and recommendations on resettlement, flood control, the environment, navigation, design and safety and economic aspects of the proposal. The panel found that lessons of the past had not been learned. Economic, human and ecological costs had been under-estimated or under-stated, while projected benefits had been exaggerated. One of the panel members, Vaclav Smil, author of several books on energy and the environment in China, wrote, 'this is not engineering and science, merely an expert prostitution, paid for by Canadian taxpayers.'[17] CIDA withdrew from the project.

China continued its plans without the aid agencies, preparing to flood 4 cities, 8 towns and 356 large villages along with vast stretches of farmland. An estimated 1.3 million people will be displaced in the process, part of what some predictions say will be a financial, social and economic disaster. Officially budgeted at $30 billion, some critics say it will exceed $75 billion. And that may not be all. Three thousand two hundred Chinese dams failed between 1949 and 1999, including two in Hunan Province in 1975, leaving hundreds of thousands of people dead. In 1999, the official *China Daily* said that the government planned to reinforce 33 000 dams and dykes that had been poorly built during the Mao years.[18] Another report said that of the 20 bridges built as part of the Three Gorges resettlement effort so far, all but 3 had quality problems and 5 had been destroyed.[19]

Despite the criticism, however, when the Chinese government released its first Three Gorges bond offering of $120 million in 1997, it sold out in five days.[20] Governmental export–import banks in Europe and North America lined up to provide export guarantees to hungry suppliers and engineering firms eager to do business on a project whose cost estimates had risen from $11 billion in 1992 to several times that by the end of the century.

Environmentalists who dislike the project often fail to mention the most evident alternative. China's coal-fired power generation will increase global carbon dioxide emissions by 17 per cent in the next 25 years unless something takes their place.[21] Those who fix their sights on carbon dioxide emissions, however, often fail to mention the Three Gorges project as an alternative, for fear of running afoul of their dam-busting confrères. In a dismal environmental tour around the future, Mark Hertsgaard devotes a full chapter of *Earth Odyssey* to the global problem of Chinese pollution from coal-fired energy, but only four sentences to the Three Gorges Dam.[22] China is dammed with the Three Gorges project, and damned without it.

Disintegrated rural development

Beginning in the late 1940s and growing throughout the 1950s, 'community development' gradually became a cornerstone of development policy and programming. The concept grew from early successes in India to become by 1960 the basis of national, regional or localized programmes in dozens of countries of Africa, Asia and Latin America. Community development included both economic and political development objectives, and was viewed as a process that involved the direct participation of people in the solution of their common problems. Community development facilitated the transfer of technology to the people of a community for solutions to common problems. The *process* in achieving the goals was an important element in the concept.

Attention shifted away from community development in the 1960s, however, when it became clear that it was not meeting the political or the economic expectations placed on it. The proceeds of community development and the institutions it created were often co-opted or hijacked by local elites, and where the process did activate communities, it worried governments that were uncomfortable with criticism from below. With the Asian crop failures of the 1960s and the advent of high-yielding seed varieties later in the decade, development agencies redirected their focus and spending towards agricultural research, extension, credit and systems for supplying seeds, fertilizer and other agricultural requirements.

Then, in the early 1970s, Robert McNamara, veering from the traditional World Bank growth track, made a number of speeches pledging to direct more aid towards the poorest. 'Increases in national income,' he told his Board of Governors in 1972, 'essential as they are, will not benefit the poor unless they reach the poor. They have not reached the poor to any significant degree in most developing countries in the past, and this in spite of historically unprecedented average rates of growth throughout the sixties.'

The concept of 'integrated rural development' gained increasing currency among bilateral and multilateral organizations during the 1970s. The idea differed from previous community development approaches in stressing the integrated nature of requirements – education, health, rural infrastructure, access to credit, fertilizer and other inputs. Integrated rural development programmes complemented a 'basic needs' approach initiated by USAID in 1973, aimed at redirecting efforts towards meeting the basic needs of the poorest. Between 1975 and 1983, annual bilateral and multilateral spending on agricultural development, in which the integrated and basic needs approaches predominated, increased from $6.9 billion to $11.6 billion (at constant prices). World Bank agricultural projects during the period 1979–83 totalled $10.9 billion.[23]

By the end of the 1980s, however, the development assistance wheel of fortune had turned again, and the concept of integrated rural development became the subject of debate and controversy. A seminal 1988 World Bank document on poverty alleviation stated the about-face with startling clarity: 'Rural development has become a bad word in the Bank's vocabulary, as well it should be, given the poor performance of rural development projects.'[24] Many projects had failed to generate expected economic returns, the report stated.

> Their organizational structure was unsustainable following the withdrawal of external funds; benefits [as with previous community development efforts] were being siphoned off by the not-so-poor farmers; they perpetuated inefficient public sector control over marketing activities; and, most importantly, extractive macro-economic and sectoral policies undermined project intent ... Many favourable project investments made under favourable commodity price conditions suddenly became unviable once world prices began to fall.'[25]

An example occurred in Thailand where, although rice exports rose by almost 330 000 tons in 1986, earnings declined by $112 million because of a drop in world prices. World prices subsequently recovered, but the dip led the World Bank and other lending institutions to halt investments in irrigation.[26] The Bank's review went further, saying that the strategy itself was flawed in that it focused on farmers with productive assets, ignoring those for whom the programmes were intended. A further criticism turned on the problem of 'scaling up'. There have been examples of successful small integrated rural projects, but few have been replicated without losing their purpose in bureaucracy and governmental control.

And many were clearly unsuccessful. A 1984 EEC Court of Auditors report studied 12 large agricultural projects in Senegal, Mali and Mauritius and concluded that not one was profitable; that they could survive only with continued external inputs and that only half could be made viable – with radical alterations in approach.[27] Similar conclusions were reached on a variety of USAID and British rural development projects. Many had become massive enclave operations, planned, managed and financed by outsiders, heavily reliant on approaches and technologies that could not be made to function efficiently, actually exacerbating the problems they sought to solve. In Sri Lanka, for example, successive waves of tractors, provided under British aid programmes and subsidized by government, had the effect of contributing to unemployment and the economic polarization of rural communities. Estimates calculated the displacement effect of one tractor at between 500 and 1000 days of work per annum in Sri Lanka.[28]

In a review of the short-term results of over a thousand projects, a 1985 World Bank study found that the failure rate in Asian agricultural projects was 5 per cent, while in West Africa it was 33 per cent and in East Africa, 50 per cent. Five years later, the longer-term results were considerably worse.[29] Commenting on fifteen years of rangeland management projects, the FAO concluded that a billion dollars had been 'largely wasted'.[30] According to the World Resources Institute, some of the reasons are straightforward:

> Aside from policy and institutional problems, cultural differences among donors and recipients may help explain project failure . . . misreading economic or social conditions was a factor in virtually every project in which new technology was rejected. The introduction of technology that was unfamiliar or required major life-style changes was a major cause of rejection.[31]

This is not to say that there should never be tractors or new technology, or that the poor should always be protected from the unfamiliar. It is to say, however, that aid agencies have a responsibility to calculate the economic and social costs and benefits of new technologies in order to ensure that they will take root in the most sustainable way possible.

The failure of integrated rural development projects is an indication of the complexity that led to the development of the concept in the first place; it is not necessarily a valid justification for the general donor retreat from rural development, especially if there was genuine concern – as the critical evaluations suggest – for the poorest. As in the 1960s, the second retreat was justified largely on the failure of the projects to achieve a reliable food surplus, when an equal, if not more important premise was *equity*. In justifying the retreat, however, some writers went so far as to suggest that growth and equity are actually antithetical.[32]

In their eagerness to achieve quick and visible results, donors pretended for decades that they knew what they were doing in rural development, and that solutions to the extraordinary challenges of hostile climates could be found on the menu of fads and fashions that they offered. It is not difficult to criticize failure. The problem, however, is not the failure; it is the exaggerated donor claims of success, the failure to learn from failure, and the constant re-invention of wheels that will not carry the load they are designed to bear.

The twenty-ninth day

Abundant examples similar to the dams and integrated rural development projects can be found in other sectors and sub-sectors: a health industry fixated on hospitals; education systems focused more on school and university buildings than on what goes on inside them; a major bias in spending on urban areas rather than on the countryside where half the world's people still live. The result has been far from satisfactory in terms of stated objectives. Unstated objectives – such as the commercial imperative, and hidden persuaders – like the world-view of bureaucrats and engineers – will be discussed in future chapters. Basically, however, the costs of a very high proportion of mega-projects have been out of all relation to their achievements. And the high hopes attached to them for development and poverty reduction have not been realized.

Now the bad news. Even if the large-scale, quick-fix, technology-bound approach so favoured by Southern governments and Northern aid agencies had worked, there are new doubts about the heavy reliance on growth, whether growth in the classical economic sense, the 'growth with equity' concept once fostered by the World Bank, or the 'sustainable growth' of the Brundtland Report. Population growth is exponential, rather than linear. A year advances in linear fashion, one day at a time. But population, the food required to feed it, the resources required to grow the food, and the side effects of the process (such as pollution), grow exponentially – by a constant percentage of the whole. It took almost 250 years for world population to double after 1650. The next doubling took a little over a century. By 1999 the process had taken only thirty years.

The problem this presents for a world with finite sources of energy and raw materials is illustrated in an old French riddle: in a certain lily pond, there is a single lily plant which doubles in size each day. On the thirtieth day it will cover the entire pond, choking all other forms of life, but in the early days it seems not to present a problem. Action to cut it back is planned for the day that it covers half the pond. What day will that be? The answer, of course, is the twenty-ninth day.[33]

Like all parables and riddles, this one has its limitations. But in underscoring the importance of conservation, economist Herman Daly makes a clear distinction between growth and development. He argues that continued exponential growth, based on the exponential depletion of resources in a finite ecosystem, simply cannot happen.[34] If it is true, as scientists tell us, that the global economy currently uses one quarter of the world's net primary product of photosynthesis, then growth by a factor of four is probably utopian. Beyond that, to the factor of between five and ten suggested in the Brundtland Report, is impossible.

This theme has been consistent in the writings of the Worldwatch Institute for years. It argues, for example, that 'a commitment to the long-term improvement in the human condition is contingent on substituting environmental sustainability for growth as the overriding goal of national economic policymaking and international development.'[35] Daly acknowledges, however, that politically it is very difficult to admit that growth, with its almost religious connotations of ultimate goodness, must be limited.

> But it is precisely the non-sustainability of growth that gives urgency to the concept of sustainable development. The earth will not tolerate the doubling of even one grain of wheat 64 times, yet in the past two centuries we have developed a culture dependent on exponential growth for its economic stability. Sustainable development is a cultural adaptation made by society as it becomes aware of the emerging necessity of non-growth ... To delude ourselves into thinking that growth is still possible and desirable if only we label it 'sustainable' or colour it 'green' will just delay the inevitable transition and make it more painful.

This is not to suggest an abandonment of the poor, nor the concept of growing incomes and increased material consumption in the South. Greater efficiencies in the use of existing resources, the diversion of expenditure from non-productive enterprise (such as arms) to productive use, the development and application of more appropriate technologies, increased rates of regeneration and the development of renewable substitutes, all offer considerable elbow room, if not optimism. The counterpoint to the growth issue is that negative rates of growth will not advance the cause of development or contribute to the alleviation of poverty. Neither will worthless currencies and double- or triple-digit rates of inflation. But if left to chance, the relationship between sustainability and development in developing countries, and the relationships between the North and the South will remain as dysfunctional as they have been for the past generation. A rising tide does not float all boats; those fixed to short anchor chains are dragged under.

Thomas Huxley observed that it is the customary fate of new truths to begin as heresies and to end as superstitions. Robert McNamara's revelation – that increases in national income will not benefit the poor unless they reach the poor – began as an economic heresy. Today it is as much a superstition as a truth, one to which more lip service than attention is paid. The new heresy is an old phrase, the logic of which is becoming more obvious: there *are* limits to growth.

CHAPTER IV
The third sector and the Third World

A government, by itself, is equally incapable of refreshing the circulation of feelings and ideas among a great people, as it is of controlling every industrial undertaking. Once it leaves the sphere of politics to launch out on this new track, it will, even without intending this, exercise an intolerable tyranny.

Alexis de Tocqueville

During the first 'Development Decade' and into the 1970s, non-profit activities in the South were generally regarded by governments, and by bilateral and multilateral development agencies, as peripheral to the development enterprise. Donor agencies regarded non-governmental organizations (NGOs) as well-meaning, but largely irrelevant; governments regarded them as unnecessary. In the late 1970s and early 1980s, however, interest in NGOs grew, reflecting a combination of factors: the growth and greater professionalism of the non-governmental, or what is sometimes called the 'third sector'; a disenchantment with the large and often unsuccessful integrated rural development projects of the 1970s; disappointment at the lack of direct and visible impact by larger agencies and Southern governments on poverty. There was also a greater emphasis among OECD member countries on the private and non-governmental sector as an aid delivery mechanism.

Historically, NGOs have actually been at the forefront of important development initiatives, including primary health-care, women's and gender issues, credit for the assetless, and participatory approaches to development. The concepts of intermediate and appropriate technology grew entirely from the non-governmental sector. NGOs are almost exclusively responsible for placing environmental issues on the development agenda. They have a greater ability than government to reach the poor, often in remote areas; they are innovative; they adapt easily to new conditions and situations; and they operate at low cost. Above all, the strength of NGOs lies not so much in paying for development projects, as in the mobilization of people into organized group action for self-reliance and self-development.

The financial contributions of NGOs should not, however, be overlooked. World Vision has become the largest international NGO in cash terms, going from strength to strength in dozens of countries and becoming by far the largest fundraising operation in the United States, Canada, Australia and New Zealand. It also has significant operations in Britain, the Netherlands, and Japan, and like other large NGOs, it has started fundraising in the South. By the late 1990s, World Vision had become an organization with global annual resources of more than half a billion dollars.[1] The annual spending of World Vision, in fact, is greater than the ODA of several OECD member countries – Finland, Ireland, Austria, Luxembourg, New Zealand, Portugal.

CARE has reached proportions similar to those of World Vision, although the bulk of its income is still US-based, and a much greater proportion of the total is in-kind contributions, mainly food aid. The combined 1997 cash and kind income of CARE members in North America, Australia, Japan and Europe was $432 million. Other 'super NGOs' include Foster Parents Plan (Plan International) with global income in 1997 of $277 million, and Oxfam, with combined international 1997 expenditure of $253 million. The total income of Save the Children Alliance members in 1998 was over $350 million in total revenue.[2] *Médecins sans Frontières* (MSF), founded in 1971 by a group of French doctors, had by the end of the century become a multi-faceted international network of health-related programmes, primarily in emergency situations. In 1996, MSF had 15 000 staff, it maintained 1131 field posts and dispatched an additional 2400 doctors and health workers, with a global income of $252 million.

There is an unflattering side to NGOs, however, characterized by opportunism, self-righteousness and amateurism. Sheldon Annis points out that 'NGOs are so frequently lost in self-admiration that they fail to see that the strengths for which they are acclaimed can also be serious weaknesses. In the face of pervasive poverty, for example, "small scale" can merely mean "insignificant". "Politically independent" can mean "powerless", or "disconnected"; "low cost" can mean "underfinanced" or "poor quality"; and "innovative" can simply mean "temporary" or "unsustainable".'[3] Self-evaluation has generally had a low priority among NGOs, in part because people make donations on the basis of an appeal, rarely asking about the quality of an organization's work. Traditionally, there has been little relationship between an NGO's performance and its survival.

Some Southern observers are increasingly concerned about the development of local NGOs on other, more basic ethical, financial and political grounds. Some charge that the professionalization and institutionalization of NGOs contradicts a traditional or religious voluntary ethic: career development workers negate the virtue and integrity of voluntary service. Others worry that the voluntary sector increasingly diverts external bilateral funding away from government plans and priorities, and that some NGOs are fronts, or could become fronts, for unwanted political activity.

The historical context

Humanistic service and the philosophy behind it is neither new, nor does it emerge from a particular place or time. At its most basic, even in primitive ancient societies, it grew from the elemental obligation of the family, tribe or clan to ensure that its members were fed, housed and clothed. Four thousand years ago the Code of Hammurabi ordered Babylonians to see that 'justice be done to widows, orphans and the poor'. Humanism, focusing on family and society rather than the individual, predominated in ancient Chinese philosophy. The Confucian code of ethics, based on goodness, benevolence and love for all, was developed in the sixth century BC and the subsequent advent of Buddhism imbued it with spiritual underpinnings that have survived 2500 years. Hinduism stressed the transitory nature of life on earth and the importance of doing good works in order to gain release from the cycle of rebirth and therefore achieve salvation: 'If a man does justice and kindness without sin, his reward is stretched into other lives which, if his virtue persists, will be reborn into a loftier place.'

Basic Jewish law, originating 3300 years ago, stressed the importance of tithing one tenth of a person's income to charity. Greek philosophy and ethics have influenced Western thinking for over 2500 years; it was in the Golden Age of Greece that the common sharing of political authority – democracy – was developed, based on a financial system of contributions rather than taxes for education, health, culture and the welfare of the aged. Christ prescribed charity as a means of salvation: 'If thou wilt be perfect, go and sell what thou hast, and give to the poor, and thou shalt have treasure in heaven.' Seven hundred years later, charity became one of the five obligatory duties of Islam through Zakat, a special tax on property and income for redistribution to the poor.

Through the Dark and Middle Ages in Europe, the Christian church became the dispenser of benevolence and charity, establishing hospitals, schools and universities, and by the twelfth century in England, it controlled half the country's entire public wealth. With the advent of Protestantism across much of Europe, a 'work ethic' developed which equated poverty with sin, and established its alleviation among the helpless as the responsibility of the state, or of parishes and villages. Among the able-bodied, poverty was basically regarded as a punishable evil, a concept resulting in work-houses, debtors' prisons and widespread misery. The Elizabethan Poor Law of 1601 codified this approach and influenced charitable thinking and deeds in North America for two centuries, and in Britain for three.

In nineteenth-century Europe, it was reformers and reform movements rather than governments who brought about changes in working conditions, public health and child labour. The pressure of these reformers, the almost uncontrolled proliferation of charitable societies, and the political threat of a potentially revolutionary underclass gradually urged greater welfare initiatives on government. Germany's pioneering social welfare programmes of 1880 were followed by France's 1893 National Law for Free Medical Assistance for the poor, and by Britain's old age pension scheme of 1908 and the National Health Insurance and Unemployment Insurance Acts of 1911.

It is in North America, however, where government was slower to respond to welfare needs, that the importance of the non-profit sector can be seen most clearly. In 1995, for example, it was estimated in the United States – where there are well over a million philanthropic and non-profit organizations – that the sector employed 13.5 million people, accounting for 12 per cent of all jobs in the country.[4] It is the societal role of the non-profit sector rather than the economic, however, which has had the greatest impact in the United States. To appreciate this societal impact, it is helpful to go back 150 years, to the 1835 visit to America of the French political writer, Alexis de Tocqueville. In his opus, *Democracy in America*, de Tocqueville commented again and again on the unique American propensity to create and join voluntary organizations:

> Americans of all ages, all stations in life, and all types of disposition are forever forming associations. They are not only commercial and industrial associations in which all take part, but others of a thousand different types – religious, moral, serious, futile, very general and very limited, immensely large and very minute. Americans combine to give fêtes, found seminaries, build churches, distribute books, and send missionaries to the antipodes. Hospitals, prisons and schools take shape in that way. Finally, if they want to proclaim a truth or propagate some feeling by the encouragement of a great example, they form an association. In every case, at the head of any new undertaking, where in France you would find the government or in England some territorial magnate, in the United States you are sure to find an Association.[5]

It was de Tocqueville's observation – less than fifty years after the independence of the United States – that these associations formed the basis of America's development and the roots of her democracy: 'In democratic countries knowledge of how to combine is the mother of all other forms of knowledge; on its progress depends that of all the others. Among laws controlling human societies there is one more precise and clearer, it seems to me, than all the others. If men are to remain civilized, or to become civilized, the art of association must develop and improve among them at the same speed as equality of condition spreads ... The more government takes the place of associations, the more will individuals lose the idea of forming associations and need the government to come to their help. This is a vicious circle of cause and effect...'

It is in international development and in the redefinition of 'the art of association' and 'community' that the non-profit sector takes on immediacy for developing countries; for until recently, most of the prominent non-governmental organizations operating in Africa, Asia and Latin America, were foreign rather than local. American foundations such as Ford and Rockefeller have played a critical role in the reduction or eradication of malaria, yellow fever, cholera and hookworm, and in the development of high-yielding varieties of wheat and rice. Organizations like Amnesty International and Human Rights Watch have acted as the world's political conscience on human rights abuse. Greenpeace, the Worldwide Fund for Nature and dozens of environmental organizations have kept nature and the environment high on the global agenda, while Oxfam, CARE, MSF and others have acted as monitors of human disaster, shouldering much of the responsibility for subsequent international relief activities.

It is arguable, however, that voluntarism does not travel well. International NGOs are an extension of the most basic elements of their donors' altruism. But originating in a foreign country, they can hardly be described as an expression of local concern – either collective or individual. They have other shortcomings: cultural insensitivities; technical and administrative weaknesses. Even those organizations that support local NGOs cannot avoid imposing their views of development, their priorities, and their bureaucratic requirements on the recipient.

There are other, more systemic problems. The vast majority of people in Europe and North America who make donations for international development want their money used for the immediate and direct alleviation of poverty. Conditioned by a generation of disaster appeals, they are much less interested in structural change, macroeconomics and vaguely ominous concepts of 'empowerment'. Torn between a short-sighted, emotional donor base and Southern realities, Northern NGOs have tended to staff themselves with generalists rather than specialists. One study observes that 'NGOs are typically rather cautious in their attitudes towards professionalism, an attitude which derives partly from their dependence on short-term fund-raising, but partly also from their pride in their value-orientation, and the desire not to impose intellectual barriers which might serve to dampen the ardour of their ordinary membership.'[6]

A typology of humanistic service

Dr Johnson observed that definitions are tricks for pedants. Perhaps that is why the fast-growing body of literature on different types of NGOs is not very helpful in this regard. In addition to definitions, there have been several attempts to rename the sector – Private Voluntary Organizations (PVO) is favoured in the United States, but there are others: Private Development Organizations, Non-Governmental Development Organizations, Public Service Contractors, the Third Sector. But the term, 'non-governmental organization' seems intent on staying, for good reason. David Brown observes that NGOs see themselves, by and large, as being what governments are not: 'not bureaucratic, not rigid, not directive and not stultifying of local initiative. This image plays an important functional role in freeing NGOs from established political hierarchies, though it does at the same time ignore the extent to which many of their projects are interdependent with governments' own activities.'[7]

The evolution of humanistic service varies in education, health and social services from country to country, over time, and according to the pace of development and the predominating ideology. David Korten has classified international NGO development into four 'generations'. First generation NGOs concentrate largely on relief and welfare. Second generation organizations focus on small-scale, self-reliant development activities, and third generation organizations concentrate on sustainable systems development. Fourth generation organizations are broad-based 'movements' such as the Chinese literacy movement of the 1920s, or the environment and women's movements of today.[8] This typology implies that organizations graduate, or should graduate, from simple or even simplistic welfare-type activities to something on a higher plane. It suggests that welfare may be a second-class activity, and it leaves little room for a welfare organization to get better at what it does.

Samuel Martin classifies the evolution in societal rather than organizational stages. Stage One is characterized by a high degree of direct personal involvement and responsibility for the delivery of humanistic service. Small, community-based self-help efforts and personal voluntary service fall into this category, which was most common, in most countries, until the early years of the twentieth century. Stage Two grows out of necessity – the institutionalization of humanistic service because of greater needs, and greater concentrations of people. Responsibility remains with individuals – but through the formation of associations. These might focus on welfare or development, or any number of things.

Stage Three evolves from convenience rather than conviction: the demand for services, the proliferation of associations and the intensification of fundraising lead to the federation and professionalization of associations, and – often reluctantly – to government funding or to replacement by government agencies and departments. And Stage Four is the ultimate humanistic delivery system, in which a society decrees that all members will be provided – by government – with an appropriate level of health care, education, cultural enjoyment and social well-being; charities and voluntary organizations become redundant and cease to exist.[9]

The stages are not immutable, however, nor is one necessarily superior to another. Time, societal values and resources will make that distinction. In fact, most of Europe in the Middle Ages – under heavy church rather than governmental domination – had reached a form of Stage Four, only to move backwards to Stages One, Two and Three through the nineteenth century, re-emerging at Stage Four after the Second World War. More conservative political ideologies in western Europe and the collapse of eastern European communist regimes which espoused Stage Four services (but provided considerably less), demonstrate a return to the Stage Three approach.

The movement from Stages One to Three can be seen more clearly in the development of Southern associations such as the Organization of Rural Associations for Progress (ORAP), which within four years of its 1980 beginnings, had linked over 300 small Zimbabwean associations and by 1998 had 1.5 million members federated through dozens of umbrella organizations and district associations. The 'Six-S' movement in Burkina Faso was founded on the ancient *kombi-naam* self-help tradition, growing by 1990 to 4000 groups of farmers – organizing cereal banks, building warehouses and small dams, and digging wells. In the Cordillera region of the Philippines, an alliance of women's organizations – *Innabuyog* – was formed in the 1980s, and within a few years had a membership of 82 self-help groups, mostly in remote rural areas, working on the development of basic services, education and human rights.

Because of their motivation and donor base, Northern NGOs tend to be highly individualistic and territorial, and often have a strong welfare orientation. Proliferating at a staggering pace in the 1970s and 1980s, those most successful at fundraising usually make their appeal to the donor's sense of personal responsibility and the direct delivery of services to those in need.

Historically, in the South, the altruistic ethic has manifested itself in a community-based approach to humanistic service, very much of the Stage One type. During the colonial era, branches of larger metropolitan charitable organizations as well as new, home-grown varieties were planted or transplanted, and often flourished. Many, especially mission-based activities, operated in the fields of education and health. Throughout much of the former colonial world, these have now been taken over by governments. This left the field open to a newer breed of more secular international NGOs, usually working in the less well-defined areas of social welfare, rural development and relief.

It is predominantly in the more recent emergence of development-oriented *Southern* NGOs, however, that one is likely to find more profound societal impact. In Sri Lanka, Sarvodaya – covering a third of the country's villages – was one of the few national institutions, including government, able to function in all parts of the country during the devastating civil war. The much-praised Aga Khan Rural Support Programme covers 80 per cent of the villages in Pakistan's Northern Areas with extension, technology transfer and credit facilities. Similar spin-off non-profit and government-supported organizations have been developed in Punjab, the North West Frontier Province and other parts of the country. The BAIF Development Research Foundation in India[10] illustrates an evolutionary process that combines sound technical competence, high ethical purpose and impact. The organization was founded by a young man – Shri Manibhai Desai – who had become a disciple of Mahatma Gandhi in 1945, working on cattle breeding for over 20 years as part of a Gandhian order. Although the organization he subsequently founded in 1967 never veered from its Gandhian roots, Manibhai put a high premium on scientific excellence, bringing in experienced technicians and ensuring that lessons were learned and remembered. Working closely with government and farmers, one of BAIF's achievements was a high-quality artificial insemination service, delivered at lower cost than government, combined with an extensive programme of progeny testing, research into semen quality and on-farm services. BAIF pioneered frozen semen technology in India and is today estimated to have been responsible for producing as much as ten per cent of the entire Indian cross-bred dairy herd.[11]

The influence of Grameen Bank is legendary, not only in Bangladesh, but in other countries such as Malawi, Sri Lanka and Malaysia, where local versions have been initiated. The Bangladesh Rural Advancement Committee (BRAC) has developed a non-formal primary education programme which was operating 33 000 schools for more than a million children by 1999. The BRAC dropout rate was only 5 per cent, compared with 80 per cent in government schools. Similar examples of quality and impact can be found in India's Self-Employed Women's Organization (SEWA), the Mysore Resettlement and Development Agency (MYRADA), Proshika in Bangladesh, the Centro de Estudios y Promoción del Desarrollo (DESCO) in Peru, the Kenya Rural Enterprise Programme (KREP) and others. For each of them, there are at least two or three more next-generation NGOs coming up that show every sign of being able to reach similar levels of coverage and effectiveness. Without pushing the analogy too far, it could be argued that these Stage Two and Stage Three organizations represent the same phenomenon – a generation or so after the independence of their countries – that de Tocqueville so admired in the United States of 1835.

Civil society

'Civil society', a neglected concept since Hegel, Gramsci and others tackled it, took on new life toward the end of the 1980s, thanks in part to ferment on the subject in both Eastern Europe and Latin America. By the end of the twentieth century, civil society discourse had mushroomed into a voluminous library of books, papers, monographs and conference proceedings. Like the definition of an NGO, definitions of 'civil society' occupy chapters and even entire books, at the end of which the average reader is often none the wiser. Charles Bahmeuller's definition is as good as any: 'The term, 'civil society' refers to voluntary social activity, activity not compelled by the state. Civil society is the whole web of spontaneous social relationships that lie outside the institutions of the political order and legal duty.'[12]

This obviously includes NGOs. When the new civil society discourse began, however, there was thought – at least in some quarters – that NGOs *were* civil society. This idea coincided with two major changes in the way official development assistance was conceived. One of these changes was heralded by the collapse of the Berlin Wall, after which many donor governments felt less constrained to support corrupt and undemocratic regimes. 'Good governance', an innocuous little tune first heard in the early 1980s, now became a full-fledged anthem. In the search for alternative delivery mechanisms in countries where 'governance' was weak, donors began to recognize the usefulness of NGOs. It was possible to reduce, or even cancel bilateral assistance to a country like pre-Aristide Haiti, for example, but through NGOs, emergency assistance and support to the most vulnerable could continue with political impunity.

The search for good governance was in some ways less important for NGOs than the enthusiasm for structural adjustment that also began sweeping through the corridors of international financial institutions at the beginning of the 1980s. Among the primary tenets of structural adjustment was the notion of less government and reduced spending on social services. Here again, NGOs (or civil society – in this scenario the terms are interchangeable) came to the fore. NGOs could fill gaps, rally communities, build social capital. Like the pieces of a jigsaw puzzle, it all fit together nicely, at least in theory: less government, better 'governance', the opportunity to reduce aid to bad regimes (and then to simply reduce aid altogether).

The upshot of the debates on civil society, structural adjustment and good governance was not negative for NGOs. In the heady donor rush actually to find civil society 'actors', or gap fillers, or whatever was required to meet a challenge, NGOs often came up trumps. Northern NGOs could assist with a range of democracy-type programmes, ranging from election monitoring to reconciliation work in an emergency situation. In helping to form and support *local* institutions, they could strengthen community efforts, build new civil society organizations, fill gaps, provide new welfare services, and so on. During the mid to late 1980s some bilateral donors began to fund Southern NGOs directly, in the belief that the Northern intermediary offered little in the way of added value, and might actually get in the way of the kind of institution, or service, or concept they had in mind. So Southern NGOs too, came up trumps.

The ink was barely dry on the first generation of civil society studies, however, when it became fashionable to actually *criticize* NGOs as *non-exemplars* of civil society. Civil society, so the new argument went, was much more than NGOs. It included the arts, sports, religious groups, political and labour associations, organizations of journalists, academics, women, the aged and youth. NGOs, it turned out, were only a *small* part of civil society. It soon became difficult to find a study on development assistance and civil society that did *not* take a swipe at NGOs for being something akin to imposters in this great panoply of democratic and societal glue.

It is perhaps useful to remember that regardless of where they fit, NGOs did not invent, or even re-invent the expression 'civil society'. Few have ever claimed that they were in the vanguard of civil society. Most do not use the expression at all. It was donors in search of a new label, one that might kill many birds with the same small pebble, who applied it first to NGOs, and as quickly began stripping it off when it no longer suited their purpose.

Proliferation, duplication and replication

A serious problem among NGOs (and for donors) is the proliferation of organizations, both in the North and the South. In 1998 there were 400 British non-governmental organizations concerned with and involved in Third World development. In the United States there were 420, in Canada 274, in Denmark almost 200 and in Australia 120. In Austria 50 NGOs received support from government, but there were another 650 that reported some interest and activity in support of international development. The comparable number in Germany was 2000.[13] Usually the multiplicity is justified on the basis of diversity, on the grounds that as community-based organizations, these organizations inform, educate and sensitize their different 'constituencies' on important issues, and create a foundation of understanding and public support for larger bilateral development programmes.

There is another way of looking at it. The proliferation, regardless of its positive attributes, is costly, selfish and sometimes destructive. Inevitably, each organization must have staff, offices, typewriters, photocopiers, telephones, meetings, newsletters, and, above all, projects. The projects necessitate overseas offices, staff and travel to developing countries, as well as some means of demonstrating to individual donors that the organization is carrying out useful and unique activities. For most Northern NGOs this means the need for 'partner' organizations in the South, many of which exist, in effect, solely at the discretion of their Northern benefactors. It means that for many NGOs – North and South – life is a project mill in which opportunities for learning, specialization and professionalism are limited.

The same multiplication exercise is now taking place in the South. Estimates placed the number of Indian NGOs in 1998 – large and small – at almost a million. In Brazil there are over 210 000 NGOs listed in government registries. The non-profit sector in these two countries is especially large, but the numbers elsewhere are also large and growing: Egypt, 17 500 in 1998; Ghana, 800 registered with the Department of Social Welfare; Thailand, 15 000.[14] As Sheikh Mujibur Rahman, the first President of Bangladesh once observed, where there are bees, there is honey. Before 1972 there were many East Pakistani social welfare organizations, but there were few, if any, that could be called development organizations. By 1998, there were 840 registered Bangladeshi organizations approved for receipt of foreign funding, and some 14 000 village-based organizations registered with the Ministry of Social Welfare.[15]

The proliferation in the South is justified in the same way as in the North: any organization that helps the poor to improve their lot is justifiable. But at what cost? It is arguable that underlying the laudable intentions and grand statements about community participation and self-help, lies a fundamental unwillingness on the part of NGOs, North and South, to share, to subsume organizational identity and ego into a larger pot for the greater good. For all the attempts at co-ordination of NGOs, for all the umbrella organizations and clearing houses, and for all the talk of NGO solidarity, genuine sharing, not to mention mergers, is the exception rather than the rule.

Like theories of social development, 'empowerment' is an intangible thing, a process open to broad, subjective interpretation. It is hard to quantify, and from the perspective of institutional donors, difficult to evaluate. Like others, the Bangladesh Rural Advancement Committee (BRAC) has struggled with questions of empowerment and replication. BRAC began in 1972 as a small, localized relief operation in Sylhet, but its workers quickly discovered that permanent solutions to the alleviation of the crushing poverty of Bangladesh's rural landless, lay in the ability of these people themselves to mobilize, manage and control local and external resources.

This approach, or rather this phraseology, is used by many, if not most NGOs, to describe what they do. 'Empowerment' has become so common, in fact, and covers such a wide variety of activities, sins and omissions, that like many phrases in the development lexicon it has become virtually meaningless.

For BRAC, the difference lay in the learning, the mistakes, the success, and finally in broadening the application. By 1999 it had grown into a multi-faceted development organization with 58 000 workers. It had organized 63 800 self-reliant groups in over 37 000 villages. Its principal health programme, the introduction of oral rehydration therapy, had ten years earlier reached an estimated 10.4 million households – more households than exist in Sweden, Norway and Denmark combined – and had then been expanded into a comprehensive child survival programme. Its savings and credit programmes had, between their experimental beginnings in 1979 and 1998, generated Tk1.7 billion (US$38 million) in savings, and BRAC had provided Tk21.6 billion (US$470 million) in credit for productive enterprise. The repayment rate was 98 per cent.[16]

What BRAC had proven was that it could do things well. It had also demonstrated that it could grow. Its formula – developed, tested and applied – worked. The landless rural poor could, largely through their own efforts, be mobilized for their own self-help. BRAC had also achieved that most elusive of development objectives: replicability. What it had not achieved, however, was the means of moving on, of withdrawing after a period of time and starting afresh in new areas. And that very fundamental problem led to the development of a unique proposal.

BRAC felt that its basic village mobilization and development package, including its conscientization, group formation, health and education activities, along with a good degree of savings and credit, could be implemented, on average, within a period of three years. It had discovered however, that without follow-up, without some sort of continued presence, group cohesion withered; lessons learned were forgotten; borrowing patterns changed and deteriorated.

This is not surprising. Given centuries of colonization, given the irreconcilable factionalism of village life, given mistrust, corruption and degrading poverty, it may be unrealistic to project any kind of withdrawal from successful activities within decades, much less years. In 1990, therefore, BRAC initiated a half-way house: after the basic mobilization is achieved, after there is a prescribed level of savings and a proven credit operation in a village, BRAC's *subsidized* activities are withdrawn, and a non-profit 'BRAC Bank' takes their place. Subsidized activities are converted into a commercially viable, *self-sustaining* operation, paid for by the villagers themselves through interest on their loans.

Bilaterals and multilaterals

In its basics, the BRAC story provides a striking reason why bilateral and multilateral donors have begun taking a greater interest in NGOs. As the aid failures of one 'Development Decade' piled on the graves of its predecessors, as the private-sector orientation became more fashionable in development thinking, eyes began to turn more frequently towards NGOs. Praise and money followed. In the 1990s, received wisdom held that there were still problems with NGO management, with replicability of NGO projects, and in co-ordination between NGO goals and broad national or sectoral development priorities. But it was recognized that the field orientation of NGOs, their flexibility and focus on community participation, local leadership and self-reliance, gave them a genuine comparative advantage in many areas critical to the development process. Their costs were lower; they worked directly with the poor; they could move in areas where angels – not to mention governments and bilateral agencies – feared to tread: war zones, politically sensitive areas, flood zones, earthquakes and drought. They could respond quickly, pioneering innovative approaches in public health, community participation, the role of women, appropriate technology and non-formal education. Almost as importantly, as Northern taxpayers and journalists grew more critical of official aid shortcomings, it was recognized that NGOs could play a vital role in the development of an informed, concerned and positive constituency in support of aid programmes at home.

Meanwhile, the bilateral agencies' discovery of Southern NGOs, which began with praise, was soon augmented with money. It seemed a logical confluence of interests. To many Southern NGOs, bilateral institutions represented a new source of untapped funding and possibly a means of reaching governments – their own and those of the donors – with a grassroots development message. From the donor point of view, Southern NGOs offered a new perspective on development, a window hitherto jealously guarded by Northern NGOs. Now, through Southern NGOs, they could undertake experimental projects or pilots that might prove instructive later in larger government-to-government undertakings. NGOs were an inexpensive, effective means of delivering official development assistance to the poorest of the poor, something that could be used to good advantage in public reports.

Multilateral agencies were not far behind the bilaterals. Using small amounts of money to advantage, UNICEF, for example, successfully promoted its 'Child Survival and Development Revolution' worldwide among NGOs. The World Bank, UNDP and others began to see NGOs as potential allies in the promotion and acceptance of new development approaches, from population planning to micro-credit. 'Used effectively', NGOs could form a critical part of strategic alliances, alliances between governments, NGOs and official development assistance agencies to help achieve key objectives. 'In fact,' stated a surprised World Bank Policy Study, 'World Bank health analysts report that nongovernment providers in many countries (Ecuador, Thailand, Zambia) offer better care than government facilities. In Nigeria and Uganda mission hospitals and clinics have medicines and other supplies when public facilities do not. In Malawi consumers walk miles past nearly free government health centres to get to mission clinics that charge many times as much.'[17]

But as relations between bilateral donors and Southern NGOs developed, problems – sometimes severe problems – emerged. One of the worst was, and remains, the endemic donor fixation on projects, combined with an almost pathological abhorrence of recurring costs. At first glance, the project approach seems reasonable and practical. In theory, both donor and recipient share an agreed list of priorities, activities and time-frames. The project begins and ends, and unless there is a second phase, the donor will expect to have initiated, catalysed or sparked a development that will then be taken up by the local authorities. The concept is clear, the terms of the agreement are straightforward, the recipient is accountable, the product can be evaluated.

This reasonable, business-like approach to getting the work done has, in actual fact, high transaction costs, is sometimes anti-developmental and is often unfair. It undermines the foundation of voluntary, non-governmental organizations; it has in more than one case brought large Southern organizations to the brink of bankruptcy; and it seriously compromises the delivery of their services to the very people the donors say they want to help. Activities cannot be turned off while an NGO awaits donor approval for a project or a second phase. People do not live their lives on a project basis: farmers cannot delay planting; sick people do not cure themselves; schools and clinics cannot close; administrators and trainers cannot be sent on unpaid vacation.

Once signed, some agreements provide for very little, or even nothing in the way of advances, which means that the NGO has actually subsidized the donor. Money arrives erratically and often late, making any sort of organizational budgeting and cash-flow planning impossible. One might assume that donors are as efficient and as understanding as they expect a recipient NGO to be. This is not so. In 1987, there was a simple bureaucratic delay at GTZ in processing a claim from the African Medical and Research Foundation (AMREF), a large Kenya-based NGO. Two months passed, and then there was a query; not a query of earth-shattering proportions, but enough to hold up not only the items in dispute, but the entire payment until it had been answered. The claim, for approximately twenty million shillings (US$1.2m), was so badly delayed that AMREF was forced to eat into all its cash reserves and for two months was barely able to meet its payroll.[18]

In the North, most NGOs have at least some arrangements with their bankers to cover such contingencies. Banks understand that NGOs occasionally suffer cash-flow difficulties, but they know that inevitably, given the volume of project agreements, government support and general public goodwill, the money will come. This is not the case in the South, where banks are much more conservative, where NGOs are less well understood, where governments place IMF-inspired limits on overdraft facilities. So it was in Kenya, where the AMREF overdraft facility was only eight million shillings – 3.8 per cent of its gross 1987 income, or enough operating funds for about 14 days. In the private sector, one of the most common causes of bankruptcy is exactly this problem – cash-flow. Even the most flourishing business will go under if it cannot meet the payroll.

The problem is not exclusive to bilateral and multilateral donors, but where large sums are involved, the cost of delay and confusion can be enormous. Reporting requirements are another major aggravation. Some donors are satisfied with semi-annual reports, but most require them quarterly; some demand them on a monthly basis. With the rare exception of the occasional donor consortium, all have different formats, all want to be treated as special. What this meant for AMREF, on 50 projects in 1987–8, was an estimated 300 different financial reports and nearly 150 narrative reports. With a project management staff of nine – limited by the non-availability of funding for institutional support costs – it is no wonder that reports were late, and that they did not always conform to a donor's standard or expectations.

The reporting problem relates fundamentally to the notion of 'accountability', a word which rolls easily off Northern tongues, as though it were a one-way concept. It could, and perhaps should be asked who holds the donor accountable for inexplicable delays, for rigid reporting requirements and sudden changes in policy, for fields unplanted, children unvaccinated, clinics closed, staff laid off.

Overheads: development on the cheap

One of the biggest problems facing NGOs in both the North and the South is the issue of administrative overheads. With the exception of the United States and Austria, all OECD member countries have established blanket percentages of their grants as a contribution to NGO overheads, generally ranging between five and ten per cent. Usually there is no restriction on the use of these funds – and a good thing too, as most go nowhere near covering the real cost of doing business.

An understanding of the 'real cost of doing business', however, is obscured by the way NGO records are presented, and this, in turn, is a function of the firmly held and reasonable public belief that charities should get as much of their income to the needy as possible. This reasonable idea, however, has turned into the unreasonable demand that 'as much as possible' should equate with 'almost everything'. In practice, it doesn't work. The most vibrant criticism of NGOs over the past two decades has centred on their lack of professionalism. NGOs are told to train more, to plan better, to monitor, report and evaluate more and better. Accounts must be professionally prepared and presented; work with partners must be handled with empathy, understanding and care; experiences must be recorded, disseminated and remembered.

This all costs money. Whether its annual income is $500 000 or $50 million, an NGO must have professional accountants. To get and keep them, it must pay market rates. To keep good field staff from leaving for higher-paying government, UN and World Bank jobs, they too must be paid reasonably. Airfares, communications and reports to donors all cost money. Finding and keeping a $50 donor also requires regular reports. These must be prepared by real people with families of their own to feed. Like administration, fundraising is not cheap. When it places an unrealistic restriction on overheads, the institutional donor does several things. It forces the NGO to cut corners on necessary and legitimate administrative costs: planning; the recruitment, selection and support of personnel; procurement and shipping; programme monitoring, reporting and evaluation; financial management and reporting. Cutting corners is not the same as observing efficiencies. All donors expect NGOs to be professional, yet this is supposed to be achievable for five or ten per cent of programme – usually an arbitrary blanket figure, regardless of whether the programme relates to emergencies, long-term development, institution-building, or savings and credit. Any cost accountant would quickly demolish the 'principle' on which such calculations are based, and any private sector CEO guilty of such amateurism would quickly find him or herself without work.

Low overheads encourage NGOs to seek activities that can produce a surplus: namely contracts. The contracting of NGOs to undertake the work of bilateral and multilateral agencies – as opposed to giving them grants – has become more prominent, in part because it has become available. But for some organizations it has become essential. For some it has become the only way to offset the real cost of doing business without applying half or more of their private donor income to overheads. A final observation on low overhead allowances: it could be regarded as unethical for institutional donors – governments, the EU, UN agencies – to take their own overheads from the original funding source – taxpayers – and then to refuse a fair contribution towards the legitimate costs of organisations that are actually carrying out the work on the ground.

Southern governments; Southern NGOs

A further problem in the growing friendship between bilateral agencies and Southern NGOs is that Southern governments have begun to see in it the tip of an iceberg, and they worry. NGOs, operating under few of the constraints faced by governments, reap high praise for their work, while seeming to pinch bilateral donor funds from under the governmental nose. Southern governments demand to be consulted, and when consulted, they often resist. If a bilateral agency funds a local NGO regardless of governmental concern, the stage for conflict is set, doing little for the advancement of civil society or good governance. Thus, in their enthusiasm to support Southern NGOs, bilateral and multilateral agencies may put the newfound objects of their affection into direct contention with their own governments. This will not strengthen NGOs, it will do the opposite.

There are three ways out of the dilemma. The first is a redefinition and reconstruction of the partnership between Northern and Southern NGOs. There are many facets to this complex and much-debated issue, some of which will be addressed in subsequent chapters. For that reason, the issue is temporarily set aside until Chapter XV. The second way out of the dilemma is to ensure that any bilateral funding for an NGO is clearly seen as an *addition* rather than an alternative to regular government-to-government programmes.

The third, and most important, is the need for Southern governments to recognize the importance of the voluntary sector to any country's social and economic development. Regardless of a government's view of its own role or that of the commercial private sector, neither is likely to meet all needs of all people. Voluntary organizations, therefore, have an important role to play in assisting, complementing, and even challenging government – in meeting needs or working in areas where costs, or economies of scale make it difficult for government to operate. They can cut through procedure-bound systems that have become paralysed in red tape. NGOs can provide special services to targeted groups – such as street children, refugees or female-headed households – that fall between the strands of an official safety net. In Karachi, for example, where services to slum areas have been beyond the means of the municipal government, the Orangi Pilot Project organized 28 000 families in a low-cost sanitation programme which, between 1981 and 1986, constructed over 80 miles of underground sewerage lines, secondary drains and 28 400 sanitary latrines, with participants investing over Rs30 million ($1.3 million) of their own cash resources.

Rather than diverting bilateral and multilateral funding from government priorities, NGOs themselves have a proven ability to attract large sums of *additional* funding for development purposes, creating jobs both directly and indirectly through the purchase of goods and services from the private sector. Obviously, those countries with a large number of NGOs, such as the Philippines or Bangladesh, stand to gain considerably more in economic terms than those where the sector is under-developed or harassed.

NGOs can be particularly important in areas of research and in the testing of new techniques, new technologies, and new approaches to old problems — such as their work in oral rehydration therapy or social forestry. BRAC's non-formal primary education programme has not only pointed out new, cost-effective ways of providing the basic education that is essential to poverty reduction, it has been instrumental in attracting large new sums of multilateral funding into the formal education system.

Perhaps as important in some ways, NGOs can play a mediating role between individuals and the impersonal institutions of government. They offer people an opportunity to *participate* in decisions that directly affect their own interests. They can act as an important brake on inappropriate development models imported by outsiders, and as an accelerator for workable new initiatives — imported or home-grown. NGOs are not an adjunct to or a conduit for development activities; in a strong society they are an indivisible part of it.

And yet, despite the obvious advantages and potential of NGOs, many governments remain openly hostile to them — at both political and technical levels. They focus exclusively on NGO weaknesses and behave as though they are directly or indirectly threatened by NGO growth and achievement, sometimes treating NGO activities as attempts to subvert government policies or as an unwelcome involvement in politics.

It may be that this attitude, less prevalent in Latin America and the Caribbean but common throughout Africa and Asia, is a consequence of the myriad small, unprofessional and unco-ordinated NGOs that have proliferated in recent years. Perhaps it is because of the widespread NGO reluctance to engage in professional and open self-evaluation. But the legislation and regulatory procedures that have been developed to deal with the perceived problem have, in general, neither halted nor corrected the phenomenon. Instead, the red tape has made life considerably more complex and costly for those doing good work; it has often directed the talent and energy available for development into much less productive palliative activities, and it has stunted the organizational, technological and intellectual development that is necessary to the advancement of any society. It may be that some governments simply fear the criticism that the very existence of an NGO seems to imply. In a weak government, the sentiment is understandable, though not particularly commendable. In strong societies with confident governments, the attitude is sadly shortsighted.

PART TWO

What we know

Life must be lived forward, but it can only be understood backwards
Søren Kierkegaard

OCT 2

Full Moon
13:50 U.T.

Sukkot

He hath a share of man's intelligence, but no share man's falsehood. - SIR WALTER SCOTT

CHAPTER V
Technology in history: lies and promises

I am not digging into such things because I think the old ways are necessarily better than the new ways, but I think there may be some of the old ways that we would be wise to look into before all knowledge of them disappears from the earth – the knowledge, and the kind of thinking that lay behind it.

Robertson Davies, *The Rebel Angels*

THE COMMON EXPRESSION 'transfer of technology' has been at the centre of one of the greatest misunderstandings in the confused lexicon of international development. It implies that technology is something neutral, something that exists outside of society. It suggests that technology can be handed across a desk, picked out of a pattern catalogue (as in 'how to make a car'), or shipped by jumbo jet from the North to the South along with some 'technical assistance' to get it going. One of the greatest failings of aid programmes and of the development strategies of many Third World countries is that development has been seen very much in these terms: as a dam or as fertilizer, as a machine or a technique or a factory that can be shipped South, as though the lessons learned over two millennia about technology simply do not apply to the modern era.

Technology is a combination of knowledge, techniques and concepts; it is tools, machines, and factories. It is engineering, but it is much more than engineering. It involves organization and processes. It has to do with agriculture, animal husbandry and health. It is often highly resource-specific. It involves people, both as individuals – creators, inventors, entrepreneurs – and as society. The cultural, historical and organizational context in which technology is developed and applied is always a factor in its success or failure. Technology is the science and the art of getting things done through the application of skills and knowledge. It is, according to the American technology historian, Edwin Layton, 'a spectrum, with ideas at one end and techniques and things at the other, with design as a middle term.'[1]

Over the centuries, technology has been 'transferred' through trade and migration, through art and through religion. It has been fostered and held back by commerce, governments and educational institutions. War has played a large part in its development and retardation. It has been bought, copied, stolen, and sometimes developed through independent invention. It has been lost – occasionally on a grand scale, as during the collapse of the Roman Empire. Sometimes it has been stimulated by something very small, such as the iron stirrup which gave Turkish and Mongol armies the ability to use both hands for shooting and, therefore, considerable superiority over Chinese and Hindu horsemen who rode with wooden stirrups, or with no stirrups at all. Sometimes the stimulus – as with the invention of a successful steam engine or commercially viable electricity – has been enormous and of readily apparent import to a wide range of applications. Although justice to the history of technology can hardly be done in a few pages, this chapter will attempt to review some of the lessons that have a bearing on modern efforts to develop and transfer technology in and to developing countries of the South.

Technology, trade and warfare

Trade is one of the oldest and most common means by which technology has migrated. The prolific land and sea trade between China and West Asia between the eighth and twelfth centuries took silk and textiles from China through India and Persia to Arabia. Techniques in hydraulic engineering – dams, canals and water wheels – developed in Roman times, were improved and further developed in Iraq and Iran after the collapse of the Western Roman Empire. They bear remarkable resemblance to similar developments around the Islamic world and China from the seventh century onwards. It is perhaps significant that Sri Lanka, which had created one of the world's greatest irrigation civilizations by the first century, was strategically located on the trading route between the Middle East and China. Paper-making technology also moved along the commercial routes from China, while Arab ships and traders spread Chinese steel technology and Indian medical knowledge as far afield as Indonesia, Madagascar and perhaps Zimbabwe.

Early European trading contacts with the Americas resulted in the rapid spread of new food crops and agricultural techniques not only back to Europe, but through Africa and Asia: yams, peanuts, potatoes, maize, corn, tomatoes, tobacco. Asian fruit and vegetables such as oranges, lemons, bananas, asparagus and artichokes made their way through Europe to the Americas. Trade in West African gold contributed to the development of the iron technology necessary for more intensive mining.

Through the ages, war has been a major stimulus to the development of technology. It is perhaps significant that the only man-made structure visible from the moon, the Great Wall of China, was built not for the glorification of God, but to keep marauding armies at bay. The development of better ploughshares was often the direct result of improved sword production. The invention of the longbow and the crossbow owe as much to warfare as to the need for food. The development of iron and then of steel was spurred as much by the demand for weaponry and armour as for agricultural equipment. Gunpowder, invented in China and used initially for fireworks, was carried to the Islamic world by the Mongols once its salutary effect on enemy armies had been noted. Thirteenth-century European armies fighting in Palestine during the crusades were quick to see the value of the technology once they too had felt its devastating effect. Gunpowder and improved building technology for fortifications were two of the technologies they took back to Europe with them. Windmills, unknown in the Middle East at the time, were something they left behind.

Gunpowder had a direct and powerful effect on the development of metallurgy, first in the development of alloys. The manufacture of bronze cannon expanded the demand for tin and copper in China, India, Turkey and Europe. In the late fifteenth and early sixteenth centuries, cannons contributed to knowledge about cast iron, first in its use in cannon balls, and then for cannons themselves. Steel, developed as early as the seventh century in Persia, remained an expensive and time-consuming technology until the eighteenth century. Before that, the best steels, used primarily for swords and muskets, were produced in Persia, India and Java; in the seventeenth century some of the best European gunmakers were actually little more than assemblers, using imported Turkish barrels.[2]

It was not only gunpowder, but the technologies spurred by gunpowder which created the 'gunpowder empires' of the fifteenth, sixteenth and seventeenth centuries – Persia, the Ottoman Empire, the consolidation of Mogal rule in India, the expansion and consolidation of the Chinese and Russian empires. Guns and advanced weaponry gave Europe the means for its mastery over Africa, the Americas and much of Asia and led to the creation of vaster empires than had been dreamed of. As Hillaire Belloc astutely observed,

> Whatever happens, we have got
> The Maxim gun, and they have not.

The arms race was not a concept confined to the twentieth century. Keeping up with the latest in military technology was as important to the small European countries and city states of the Middle Ages as it was in the nineteenth and twentieth centuries. Thus technological advancement in repeating rifles and machine guns put an end to muskets. The internal combustion engine, installed inside an armour-plated box and outfitted with wheels on treads, put an end to the cavalry. The aircraft, still a novelty at the outset of the First World War and used largely for reconnaissance purposes, had within four years become a sophisticated killing machine, its scope and potential radically altered almost overnight. Rockets, first used as weapons by the Chinese in 1150, received their most important boost in the wartime laboratories of the Third Reich. The principle of jet propulsion, applied (without success) on the advice of Benjamin Franklin to an early American steamboat, became commercially viable only after the intensive experiments of the Second World War. Although the space travel and satellite technologies of our times have had enormous peaceful benefits, the research and development behind them owe more to the Cold War and to military budgets than to the peaceful march of progress.

The stimulus effect

War has obviously been one of the greatest stimuli to the development and movement of technology, but there have been others – the invention of paper, the printing press, the magnetic compass. The development of the Portuguese caravel, which could sail efficiently against prevailing winds (and thus get home from West Africa), was as important to the realization of Portuguese imperial ambitions as the development of advanced weapons.

Resources, in both their abundance and their shortage, play an important role in the process of technology development and diffusion. The connection between increasing shortages of fuelwood in eighteenth-century Europe, and the development of the steam engine is not coincidental. More costly fuelwood had a direct bearing on the need for an efficient steam engine, because steam-powered engines allowed water to be pumped from ever-deeper mines – from which coal, the substitute for wood, was extracted. This new abundance of cheap coal and the development of coke made possible better and cheaper iron smelting. The growing shortage of wood had other effects, particularly on the movement of shipbuilding technology away from Europe. By 1774, one third of all British ships were being built in North America, and following the American Revolution, much of the industry was shifted to shipyards in Bombay and to the Hooghly River south of Calcutta.[3]

The development of the railway was a powerful stimulus to the British iron and steel industries, which for a generation exported engines and rails to America, and for two or three generations to Canada, Russia, China and India. The very technologies that became available because of the shortage of fuelwood led directly to the end of the wooden sailing vessel.

The 'small' innovation can have as fundamental an impact as the large. New techniques in materials development and new designs have revived interest in propeller-driven aircraft. Better shapes and different propeller lengths, combined with new materials, led to the development in the 1980s of propeller-driven aircraft that are 40 per cent more fuel-efficient than jets.

Improvements in rail beds, tracks, engines and design have given new impetus to the railway in France and Japan. New designs of windmill with 'eggbeater' turbines and blades on a vertical axis, capable of taking wind from all directions, have given wind power a new lease on life. Like the eighteenth-century shortage of fuelwood that provided the stimulus for coal and steam in Europe, these modern innovations have their origin in the changing costs of fossil fuel.

The clock, according to the American sociologist, Lewis Mumford, not the steam engine, is the key machine of the modern era. It 'symbolizes the inventiveness of the age,' and has in all ages marked 'a perfection towards which other machines aspire.'[4] The development of the clock had its origin in the studies of astronomy that so fascinated the Greeks, the Chinese, the Indians and the builders of Stonehenge. Su Sung's great astronomical clock in China's Hunan province, commissioned in 1090, was powered by a giant water wheel with 36 buckets. At the same time clocks were being developed in India. The first European clocks emerged a century and a half later, and by the end of the fourteenth century, clockwork had become a relatively sophisticated art. Instruments that showed the position of planets in addition to telling time had been developed as the result of a combination of complex mechanical and mathematical skills. Careful gearing and measurement, the use of bearings and pinions, fine metalwork and precision balance were all essential to the building of a successful clock.

Although European clocks taken to China in the early seventeenth century were dismissed as useless – both the clock and clockwork technology having been lost in the intervening years – clockmakers in Europe had already provided a major stimulus to mathematics, experimental science and reason. One of the greatest technological problems during the age of exploration was the lack of a practical method of determining longitude, a problem which caused every great navigator – sooner or later – to become lost. The problem was solved by an English clockmaker, John Harrison, who spent a lifetime developing a seagoing clock so accurate, that in its 1762 trials it lost only five seconds after 81 days at sea.[5]

Technology historian Joel Mokyr ascribes Britain's capacity for an industrial revolution, in part, to its advanced clock and watchmaking technologies. Watchmakers in Coventry later formed the nucleus of the bicycle industry, which itself became the cornerstone of the British automobile industry. In the United States, clockmakers were among the first to put the concept of interchangeable parts into operation, and with the advent of steamboats and trains running to preset schedules, the clock became essential to the shaping of regular routines. It is perhaps no coincidence that Joseph Saxton, developer of an early gear-cutting machine; John Fitch who developed the first (temporarily) successful steamboat; Matthias Baldwin who built one of the first American railway locomotives; John Kay, who helped Arkwright develop the Water Frame; Joseph Brown, the inventor of the sewing machine, and Henry Ford all started their careers as clockmakers or clock repairers.

Sewing machines and typewriters have similarly complex histories, although more compressed. The first patent for a typewriter was granted in Britain in 1714, but the first commercial machine was introduced in 1873 by the American company, E. Remington and Sons. Remington, a manufacturer of firearms, had the machines and machinists necessary for precision manufacturing of very small parts. In order to keep the typewriter keys from jamming, Remington developed the QWERTY keyboard, scattering the most frequently used vowels and consonants over three rows, and deliberately giving 60 per cent of the work to the left hand, which moved more slowly than the right. There must be a streak of sentimentality in technology, because 125 years later, the QWERTY keyboard survives, even on computers, long after typewriters stopped jamming, and long after the use of keys ceased to have any mechanical significance.

Technology lost, technology remembered

Technology, of course, does not always move in a forward direction. It can be ignored, lost, forgotten or destroyed. Sometimes old, forgotten technologies are remembered and put to good use again. The Chinese simply lost interest in clocks and many of their other technological advances. During the 'Great Withdrawal' of the fifteenth century, China's highly advanced technologies in irrigation and agriculture, ploughs, seed drills and harrows, the development of blast furnaces, textiles, water power and maritime technology all stagnated, atrophied or were simply forgotten. As historian Daniel Boorstin puts it, 'Fully equipped with the technology, the intelligence, and the natural resources to become discoverers, the Chinese doomed themselves to be the discovered.'[6]

Technologies are sometimes deliberately extinguished. In its early days, General Motors calculated that its greatest potential competition was not from other automobile manufacturers, but from other forms of transportation. As a result, in the 1920s, in conjunction with Standard Oil of California and Firestone Tire (which had similar interests), General Motors began to buy up and dismantle electric transit companies across America. Some 44 systems in 16 states were converted to the use of diesel-powered buses and then sold with conditionality clauses which precluded the future purchase of any equipment that did not run on petroleum products.[7]

As new technologies and resources emerge, old ones often die out. When the Portuguese first arrived in West Africa, they were amazed at the quality of brass casting they found, and refused to believe that the brilliant art work of Ife and Benin had not been influenced by ancient Greek or Roman travellers. The lost-wax method of brass casting is an old one, known in many societies. But such was the quality of work in West Africa that the Portuguese took African craftsmen with them to introduce and expand the technology in Brazil.

There are sites in West Africa with evidence of an iron industry dating back as far as 500 B.C., coinciding with the development of the iron age in Britain. The earliest sites in Ghana date from the second century A.D., and by the fifteenth century, iron had become a common village industry there, producing a range of agricultural implements and weapons. When fifteenth-century European traders arrived in search of gold, this gave a spur to the iron industry, as new tools were needed for mining. But as the European taste for gold increased and as slaves were added to the list of exports, more and more textiles, agricultural implements, tools and firearms were imported. Under the influence of these imports, iron production gradually declined, and by the early twentieth century survived only in remote areas of northern Ghana, little more than a crude cottage industry. Iron technology was lost, but ironically the lost-wax method of bronze casting, which in most other societies disappeared, has survived in West Africa. No longer an essential or important technology, it continues largely through handicraft sales. The greatness of the past age is sometimes recalled, however, in the auction rooms of Europe: in 1989 a single artistic Benin bronze realized £1.3 million at Christies.

History records several cataclysmic instances of massive technology loss. The most prominent European example was the collapse of Rome and the disappearance of most of its knowledge under the might of invaders to whom the hard-won technologies of previous centuries meant little. The Mongol invasion of China in the thirteenth century cut short an unprecedented era of intellectual, artistic and technological development. The Mongol invasions of Iran and Iraq had a similar effect. The great libraries of Baghdad were destroyed and the country's extensive irrigation and hydraulic systems were laid waste. Two other examples of 'deindustrialization' are relevant. The first coincides with the European arrival in what is today 'Latin America'. Within only 29 years of the first landfall of Columbus in the new world, Spanish armies under the command of Cortez – driven by a lust for gold and a fanatic desire to impose Christianity on everyone they encountered – destroyed the Aztec armies of Montezuma and their magnificent capital, which Cortez himself had referred to as 'the most beautiful city in the world'.[8] Over the next 50 years, 25 million Aztecs – perhaps 90 per cent of the population – died from devastating epidemics of European measles and smallpox.

The Inca of Peru, noteworthy for the organization they applied to society, were next. The Inca were proficient in mining and metallurgy – in copper, gold, silver and tin. Their architecture and construction techniques were of a high order; their textile technology and their irrigation and agricultural techniques were in many ways superior to those of the invader. But nine years after the destruction of Montezuma's great city, Francisco Pizarro inflicted the same fate on the Inca ruler, Atahualpa. Within a decade, a once proud civilization lay in ruins, its people working as slaves in the gold and silver mines that would fuel expanding European empires.

India is a more particular case of deindustrialization, because after the consolidation of British rule, industrial development was diverted from an evolutionary path to one aimed solely at serving British interests. The destruction of the Indian textile industry is a well-known example. Another is the shipbuilding industry. As long as it suited the Admiralty, particularly after the loss of American shipyards, shipbuilding in Bombay and eastern India thrived. The industry did not provide the stimulus that it did in other countries, however, because when strategic and technological interests changed, shipbuilding was allowed to collapse. Skills that might have emanated from the shipyards were simply never developed. The same was true of railway development which, in Europe and America, provided an important stimulus to other industries. Although a vast network of Indian rail lines had been established by the end of the nineteenth century, the rails themselves were imported and engines were assembled from components made in Britain. As Pacey notes, 'Indian participation in any skilled role was excluded. India was forced to conform to the concept of 'transfer of technology' in its crudest form, in which one party is seen as a passive borrower of techniques, not a participant in dialogue.'[9]

A more modern example of the push and pull of technology in India can be found in the South Indian State of Kerala, where the traditional or artisanal fishing industry represents one quarter of the entire country's marine fish production. In the early 1950s, a Norwegian aid project was initiated in order to assist in the mechanization of traditional fishing boats, one of which, the *kattumarram*, has given its name to the modern catamaran. When initial motorization tests in Norway failed, attention turned elsewhere, reflecting the Government of India's view that the fishery sector was in any case, 'largely of a primitive character, carried on by ignorant, unorganized and ill-equipped fishermen. Their techniques are rudimentary, their tackle elementary, their capital equipment slight and inefficient.'[10]

The Norwegians moved on to larger boats, and using bottom trawling techniques, discovered that the waters of Kerala were one of the world's richest prawn grounds. By 1962, with prawns selling to Japanese buyers for Rs8900 per ton and fish selling for only Rs150, private investment in trawlers had become enormous. But the trawlers caught more than prawns. They disturbed fish breeding grounds and netted enormous amounts of fish, most of which were destroyed in the process. During the 1970s and 1980s, the traditional fishing industry, which employed over 90 000 people, went into serious decline. Catches of oil sardine and mackerel dropped from a peak of 250 000 tons in 1968 to only 87 000 tons in 1980.

In an effort to remain competitive, fishermen took matters into their own hands, and began purchasing outboard engines with state government loans. This provided only temporary respite, however, as their boats were not suited to the vibration of the motors, and the engines were unsuited to the locally available kerosene, which was intended mainly for lamps and cooking. When engines failed, income halted and debts mounted. The Intermediate Technology Development Group (ITDG) became involved with the fishermen at about this time, and assisted in the development and local construction of a more suitable plywood boat. Plywood was a versatile material which provided the fishermen with a cheap alternative to their traditional craft, for which timber was becoming increasingly scarce and expensive. ITDG also became involved in the testing and development of engines better suited to the rigours of Kerala. And the fishermen themselves organized a strong political lobby which sought to have trawling banned in shallower waters and during the fish-breeding monsoon season.

In addition, they experimented with a technology of their own, known in Trivandrum for centuries. In order to draw fish closer to shore, fishermen had traditionally dumped rocks with attached coconut fronds into the sea. People also remembered that a shipwreck, discovered in about 50 metres of water north of Trivandrum in 1949, had become a rich fishing reef. So, in response to the problems of the 1960s and 1970s, the fishermen began to construct artificial reefs of their own, using stones, tyres, concrete rings and anchors.

The impact of change in a fragile natural eco-system is complex, and can only be judged over time. The preliminary evidence, however, was positive. Better boats allowed fishermen to become more competitive, while the partial success of their political lobbying and the development of more than 20 artificial reefs during the 1980s paid off in terms of fishing yields – which in 1988 and 1989 and through the 1990s had returned to the levels of the early 1970s.[11] One external technology had changed life forever along the Kerala coast, while two others helped poor families cope with the damage. But perhaps the most important lesson was that the fishermen were not helpless, nor were they 'ignorant and unorganized'. With scientists now studying the effects of their artificial reefs, an old indigenous technology may have an important contribution to make in finding a solution to the sustainability of livelihoods and the Kerala fishery.

Technology and organization

Warfare has contributed immensely to another important facet of the development and diffusion of technology: organization. New technologies in the Middle Ages demanded that military organization extend beyond short-term battlefield strategies into permanent government structures and systems. The manufacture of new weapons in sufficient quantities could not be left to happenstance; military security required organized production and government support.

Improvements in technology both encouraged and flowed from better organization. The classical world of Greece and Rome, for example, used animal traction inefficiently. It was not until the eleventh century in Europe that people discovered that a horse could pull a load of more than half a ton. Greek and Roman horses, depicted in surviving statuary and art as regal animals with their heads held high, were in fact constrained from pulling heavy loads by a harness that strangled them if they put their heads down. The development of the horse collar proved that a horse could pull as much as an ox and, more importantly, that it could do so at twice the speed.[12] Combined with improved crop rotation techniques and military iron technology – now used for horseshoes and agricultural implements – vast new tracts of virgin land could be brought under the plough, accompanied, of necessity, by the development of organized estate farming. Larger farms, in turn, both justified and required greater investments in infrastructure, particularly irrigation.

The development of textile machinery was important to the British Industrial Revolution, but a more fundamental factor was the emergence of the 'modern' textile factory after 1760. These factories were made possible by the advent of steam power to drive machinery, but they were as much the product of new concepts of work: in a factory, work could be subdivided, routines could be devised and work properly supervised.

The concept of interchangeable parts, developed during the first half of the nineteenth century in Conneticut's Springfield Armory in connection with the manufacture of weapons, further revolutionized concepts of production and organization. Hitherto, each gun had been made by a single craftsman and no gun was exactly the same as any other. If a firing mechanism malfunctioned, a craftsman was needed to repair it. The concept was not universally appreciated until the early years of the twentieth century, however, when the Cadillac automobile company introduced the technique to the manufacture of horseless carriages. In 1908, under strict supervision, three randomly selected Cadillacs were disassembled, their parts mixed, and then reassembled. To the amazement of all, each one was able to drive 500 miles without incident.

The idea of replacement parts, each of which would be identical to thousands of others, not only altered the face of manufacturing forever, it changed the role of the maker from craftsman to machine operator. Before the concept of interchangeable parts could be properly developed, however, a number of preconditions were necessary: the development of jigs and fixtures; the availability of cheap and abundant steel; precision grinding, cutting and drilling tools that could perform the same function over and over without variation; a cheap and accurate micrometer.

The subsequent capacity for mass production led to refinements in techniques of organization. In early factories, machines were organized by function: all the drills were in one part of the shop, all the lathes in another, all the grinders elsewhere. Raw materials and work in progress were moved about in batches from section to section. The result was often chaotic, as a batch of parts returned two or three or four times to the drilling section, either waiting for other batches to be finished, or holding them up. New concepts of flow led to a reorganization of the factory so that batches of parts moved along a line of machines placed in sequential order. If drilling was required at three stages of the operation, with cutting and grinding between, the appropriate number of drills would be located before, between and after the grinders and cutters. From this type of 'flowline' to continuous flow was a logical next step, widely adopted by the 1880s: parts no longer moved in batches. Rather they moved one by one, significantly reducing the time that material lay idle.

'Scientific management', a concept that began to develop towards the end of the nineteenth century, broke tasks into their component parts and organized them in the most efficient way. It was an effort to overcome worker resistance to the loss of creativity, and – not least – to deal with the increasingly real risks of social upheaval and revolution. Bigger and bigger became better and better (that is, more and more profitable), once corporations recognized the importance of what David Noble calls 'the task of transforming America in a sober and scientific manner ... to analyze, rationalize, systematize and coordinate the entire 'social mechanism', to translate haphazard, uncertain and disruptive social forces into manageable problems for efficient administration.'[13] Such a translation involved economic and social research, market analysis, industrial and governmental planning, labour-management co-operation and ever more streamlined production management.

Combining the operations of two or more machines into a single machine further speeded up the flow-through, but it was Ford engineers, working on the Model T in the summer of 1913, who truly revolutionized the flowline principle. Pulling an automobile chassis gradually along the factory floor while stationary technicians worked on it, the time required to produce a car was reduced from 13 hours to less than 6. A conveyor belt installed the following year reduced overall costs dramatically, and enabled Ford to slash the price of the Model T from $950 in 1909 to $360 by 1916. Not surprisingly, the annual production of cars expanded during the same period – from 12 000 to 577 000.[14]

While European and American manufacturers relied on the consolidation of small firms into giant corporations, and on lessons learned in the 1920s and 1930s, organizational concepts were further refined in Japan in the 1970s and 1980s. Inventories were reduced to absolute minimums using the 'just in time' or 'right on time' concept: raw materials, components and finished goods arrived at each processing stage or from carefully integrated sub-contractors at precisely the right moment. Stocks did not accumulate; components did not back up. Departments – from design to market – were integrated with a care and precision unknown in Europe and America. Customer opinion and preference were sought with the determination of a salmon heading upstream to spawn. Quality, innovation and flexibility made deep inroads on the single-minded Western industrial notion of 'getting the price right.'

The outcome can be seen in the experiences of Austin and Nissan. In 1952, Austin provided technical assistance and designs which allowed Nissan to produce the A40 in Japan. Although Japanese production of the A40 declined over the years of the agreement, as late as 1959, Nissan's own Datsun Cedric was basically just an Austin clone. By 1980, things had radically changed. Japan was producing 11 million vehicles annually, while the US produced only 8 million and Britain had fallen to 1.3 million. In 1983 Nissan became the world's third largest car producer after General Motors and Toyota.[15] By 1989 in Britain, the Austin marque survived on only two models, one of them designed 30 years earlier. That year Austin disappeared into Rover, which itself eventually fell to the German car maker, BMW. There are several lessons in all this, but the primary one is that technology is more than machines. Technological success owes as much to technique, design, good planning and effective organization as it does to engineering.

Technology and institutions

Organization does not exist in a vacuum. It develops within the context of institutions: religious institutions, educational institutions, government, industry, trade unions, the modern corporation. Buddhist pilgrims travelling between India and China in the sixth and seventh centuries were carriers of vast amounts of information – not least through the written word – about such things as agriculture, chemistry, paper making and engineering. In Japan, the casting of giant bronze statues of the Buddha was both a result of, and a stimulus to, new techniques in metallurgy. Both Islam and Christianity were responsible for generations of advances in architecture and building technologies. Norman architects and masons, for example, were at work on English churches well before the Norman invasion. And eight centuries after their construction, many European Cathedrals remain the largest, most solid and most beautiful buildings in the towns and cities where they were built.

In an often brutish world, religious institutions were also places where art, another key aspect in the development and movement of technology, could flourish. Writing, the illumination of scripture and eventually printing – block printing in China, and the printing press in Europe – had strong roots in Buddhism and in Christianity.

The importance of government in the advancement of technology cannot be over-estimated. The role of the state is obvious in one of technology's greatest stimuli – warfare and the development of weapons. Government's hand (and money) can be found in technologies and techniques ranging from the development of gunpowder to the organization of interchangeable parts at the Springfield Armory. In their day, governments backed Bartholomeu Dias, Christopher Columbus and Vasco da Gama, just as they later supported manned flight, space travel, and all modern communications technology – from the telegraph to fibre optics and the Internet.

Governments have played a major role in the development of shipbuilding through the ages, and thus of exploration and trade. The obverse of encouragement can also be found: the Chinese government's decision in the early fifteenth century to *curtail* shipbuilding and to withdraw the Chinese fleet from further trade with India, Arabia and the east coast of Africa had a profound and positive effect on the development of European exploration in the next few decades.

Governments protect new technologies with a framework of patent legislation and through the erection of tariffs against competition. Richard Arkwright is remembered for the invention of the 'water frame' which helped to revolutionize the British textile industry. He is less well remembered for his successful efforts to have the government reduce its excise tax on British-made textiles. Although inferior in quality to calico imported from India, the price advantage this gave British calico resulted in a two-thirds drop in Indian imports in only eight years.[16]

Education has also been important to the development and spread of technology. The great universities of Baghdad, Toledo, Paris, Oxford and Cambridge were, of course, famous seats of learning. But technical education was, until the nineteenth century, something that took place largely outside the formal education system. Skilled workmen learned their profession through the apprenticeship system. It was with the movement away from craft-based industry towards mass production that higher education began to play an increasingly important role in industrial education, and in the production and reproduction of the corporate engineer.

Endowed by industrialists to develop scientific and management studies as early as 1847, Harvard did not actually graduate its first engineer until 1854, and by 1892 had produced only 155.[17] At Yale, agricultural chemistry and other practical subjects were taught in the attic of the chapel until 1860. It was this inability or reluctance of established American universities to go beyond polite learning and to recognize the need for advanced scientific and industrial education that led to the formation of the Massachusetts Institute of Technology in 1861. Over the following century, the relationship between American universities and industry became a close one. Business and industry worked with universities in the establishment of research facilities. Universities received much of their core funding from industry. Professors worked with business and industry in the development of curricula; industry recruited graduates straight from the campus; co-operative courses of study, combined with work in industry, were adopted by dozens of universities and hundreds of companies after the turn of the century.

David Noble echoes Thorstein Veblen in questioning what had become a love affair and then a marriage between higher education and industry:

> Veblen understood that the 'free pursuit of knowledge,' which he considered the 'end for which a university is maintained,' was being thwarted, and its radical potential checked, because it was being incorporated within the larger closed framework of big-business enterprise. The 'pursuit of knowledge' would gain considerable social stature and financial support as a result of the new industry–education co-operation, but it would no longer be really 'free'.[18]

Freedom of inquiry might be maintained and guarded, but it would become a 'controlled freedom,', a 'disciplined initiative' operating in support of, rather than against the industrial system. Quelling and co-opting opposition to the industrial system was, understandably enough, a concern very much in the forefront of industrial, if not university thinking during the first quarter of the twentieth century.

The advent of mass production, the unquestioned importance of economies of scale, the apparent inevitability of giantism and the services of its handmaiden, higher education, have exercised thinkers and philosophers throughout the century. In France, Saint-Simon predicted a society run by engineers and scientists, while others saw the beginnings of a new form of government: not democracy or bureaucracy, but *technocracy*. In the 1960s, Schumacher questioned the 'idolatry of large' partly on economic grounds, but as much on social grounds.

By the 1980s and 1990s, however, the idolatry of large was on the wane in much professional management theory – on economic and management grounds as well as social grounds. The arrival on the docks of Europe and North America of better, cheaper and *different* Japanese goods – cars, cameras, electronics, steel, biotechnology, pianos, organs and just about everything else – not to mention the presence of Japanese investors in Western corporate boardrooms, had a profound effect on forcing a reconsideration of technology and how it was organized.

Technology and the individual

The study of history has been much preoccupied with kings and presidents, with wars and revolutions and plagues. Major artists and explorers are remembered, as are some individuals credited with a major invention. But ordinary farmers, seamen, craftsmen and women, entrepreneurs, and the people David Freeman Hawke calls 'the dirty fingernail people' are too often forgotten.

Johann Gutenberg is remembered as the inventor of printing, but in fact various forms of printing had been known in Europe before his birth. Gutenberg was a goldsmith and metal caster whose critical invention was a mould for casting large numbers of interchangeable pieces of type. Robert Fulton is credited with the invention of the steamboat in 1807. In fact the maiden voyage of Fulton's boat, *The Claremont*, took place a full seventeen years after scheduled steamboat service had been introduced (and then abandoned) on the Delaware River. Fulton had combined the improved technology of two decades – better brass fittings, an improved copper boiler, significantly better tolerances, greater knowledge of water resistance – with the experience of other steamboat innovators. He also worked within a much improved economic context for steam transport. Encapsulating the essence of technology 'transfer', his response to critics who accused him of stealing from others was, 'Every artist who invents a new and useful machine, must choose it of known parts of *other* machines. I made use of all these parts to express ideas of a whole combination, new in mechanics, producing a new and desired effect, giving them their powers and proportions indispensable to their present success in constructing steamboats.'[19] Fulton is remembered; the hundreds of others who contributed to his success are not.

Samuel Colt, whose name is synonymous with small arms, conceived of the revolver himself, but had every one of his models produced for him by others. After several failures and bankruptcies, he made a fortune through shrewd government contracts, astute arrangements with investors, and high quality products made by genuine tool and machinery makers. A generation later, Samuel F.B. Morse – a landscape artist – gained credit for 'inventing' the telegraph. In fact Morse's genius was not so much in taking other people's inventions to their logical conclusion as it was in raising funds to promote the use he saw for them. Henry Ford's famous assembly line was simply an improvement on already advanced thinking about the industrial flowline.

From these examples and the many others of schoolbook history – Watt, Arkwright, Hargreaves, Faraday, Bell, Edison – it is not hard to understand why the notion of invention without context and, therefore, of a portability and disconnectedness in technology, has gained currency among aid agencies and those concerned with enhancing the pace of development in the Third World.

Although the remembered 'inventors' followed and worked with a legion of unknown craftsmen and innovators, they do stand out as what today's management literature calls *champions*. Peters and Waterman and others have written at length about the importance of champions, both within and outside formal structures of business, government and industry.[20] They describe the product champion as a zealot or fanatic – not necessarily an intellectual giant, but a pragmatist, possibly even an 'idea thief' who develops a concept 'and bullheadedly pushes it to fruition'.[21] Champions do not always succeed; in fact most of them fail most of the time. But if success is normally achieved one time out of ten, it therefore stands to reason that ten attempts and nine failures must be experienced in order to achieve the one success. Or, as Damon Runyan put it, 'All life is six to five against'.

Success is almost always the result of repeated attempts and experiments, and of a climate which encourages experimentation, a climate, most importantly, which understands and tolerates failure. Champions do not emerge automatically. Peters and Waterman argue that 'they emerge because history and numerous supporters encourage them to, nurture them through trying times, celebrate their successes, and nurse them through occasional failures. But given the supports, the would-be champion population turns out to be enormous, certainly not limited to a handful of creative marvels.'[22]

Technology moves when people move. Nineteenth-century American manufacturers were notorious for rapacious industrial espionage, sending spies to Britain, bribing technicians to emigrate, smuggling machines, particularly textile machinery, out of the country. But it was the waves of immigrants from Europe to North America bringing new skills, new ideas, and new ways of approaching problems, who had the most profound general human effect on American technology development over the previous two centuries. There is a double irony in this phenomenon for today's developing countries. First, there is little permanent migration today from North to South, or (with the exception of refugees) between Southern countries. This means that the movement of technical expertise to the South is a historical hybrid, consisting mainly of students returning from studies abroad, or of short term technical experts, usually on aid contracts.

The second irony is that attempts to develop a resource base of engineers, machinists and technicians in the South are severely weakened by emigration to the North. In 1970, UNCTAD estimated that some Third World countries were losing between 20 and 70 per cent of their annual output of doctors to the North. It was calculated that the value of imported scientists, engineers and doctors to the United States in 1970 was $3.7 billion, more than the US spent that year on all development assistance to all countries. A 1991 study showed that the trend has not abated. It is estimated that between 1967 and 1985, India transferred $51 billion to the United States in the form of skilled 'human capital' — trained scientists, doctors and engineers. Between 1974 and 1988, the number of immigrant scientists and engineers in the United States doubled, from 5.8 per cent to 10.5 per cent of the total. The five leading sources of the talent were India, Britain, Taiwan, Poland and China.[23] The further bitter irony in this is that the very embassies that process the Southbound flow of temporary technical experts on aid contracts, also process the *permanent* Northward emigration of a country's most technically qualified people.

It is in studying the role of individuals that some of the most salient lessons about the development and diffusion of technology emerge. Technological development is not automatic. It is the product of conscious and unconscious choices and decisions, all of which take place within a social and historical context. The notion that technology can be 'transferred' without modification has few historical precedents to support it. The 'transfer' will almost always involve modifications. And the process of modifying a technology will, in itself, often stimulate other developments. But the implications are clear: in adapting a technology to local needs and conditions, local skills must be available. There must be a supply of 'dirty fingernail people', thinking people who have an intimate knowledge of the material with which they are working. This is one of the most fundamental historical lessons of technology development.

CHAPTER VI
Small is beautiful

There is only one way by which poverty in the developing countries can be attacked successfully, and that is by producing more in those nations. In no one of these countries can human needs be satisfied by the simple redistribution of existing income and wealth. In these countries, small is not beautiful.[1]

Robert S. McNamara, President, World Bank

ON A LATE summer's day in 1977, when snow had already fallen in the high Alpine meadows, a train made an unscheduled stop in a small Swiss town. The large man the ambulance attendants carried off the train on a stretcher was followed by a railway official carrying a suitcase and a broad-brimmed hat. At the village hospital it was confirmed that the man, Ernst Friedrich Schumacher – 'Fritz' to his friends – had died of a massive heart attack.

German born, Schumacher was largely unknown in the country of his birth. Educated at Oxford and Columbia and interned as an enemy alien in Britain during the war, he subsequently became one of Britain's top economic planners. *The Times*, for which he had once been chief economic leader writer, published three consecutive obituaries of Schumacher, and the requiem celebrated for him in Westminster Cathedral was attended by leaders of government, the academic world and industry. Born a Protestant, Schumacher was profoundly influenced by Buddhism, yet died a Catholic. He was employed for much of his working life by one the largest enterprises in Europe (the British Coal Board) but he is best remembered for a single book – about smallness. *Small is Beautiful*, first published in 1973 and translated into twelve languages, has sold over four million copies and remains a best-seller. A friend and confidant of the rich and powerful, Schumacher was honoured by Queen Elizabeth II and consulted by Julius Nyerere, Pandit Nehru and Kenneth Kaunda. He took tea with Jimmy Carter in the White House. Yet his greatest achievements were almost entirely on behalf of the poor.

Intermediate technology: the concept

Schumacher's tryst with smallness and with greatness began unexceptionably in 1955, when he was 44, during a UN consultancy to Burma. Charged with advising the government on economic planning and fiscal policy, he was struck by the richness of Burmese culture, by the Buddhist religion and by the friendliness and dignity of people who were, in normal economic parlance, simply poor. He began to see that the 'economics of materialism', with which his professional life had been preoccupied, had its limitations and his report, focusing on the importance of rural development, spoke of the need to develop local expertise and self-reliance. It was the genesis of a professional conversion that would preoccupy him for the next 22 years.

Five years later, in a 1960 article for the *Observer*, he wrote:

> The West can indeed help the others [i.e. 'underdeveloped nations'], as the rich can always help the poor. But it is not an easy matter, expressible in terms of money alone. It demands a deep respect for the indigenous culture of those that are to be helped – maybe even a deeper respect than is possessed by many of them themselves. Above all ... it must be based on a clear understanding that the present situation of mankind demands the evolution of a non-violent way of political and economic life.[2]

Invited to participate in a series of seminars in India, he came into contact with some of the country's leading Gandhians, and in 1961 began to articulate his first ideas on intermediate technology:

> Economic development is obviously impossible without the introduction of 'better methods', 'higher technology', 'improved equipment' ... but all development, like all learning, is like a process of stretching. If you attempt to stretch too much, you get a rupture instead of a stretch ... The only hope, I should hold, lies in a broadly based, decentralized crusade to support and improve the productive efforts of the people as they are struggling for their livelihoods now. Find out what they are doing and help them to do it better. Study their needs and help them to help themselves.[3]

Invited back to India in 1962 by the government's Planning Commission to advise on economic development policies, Schumacher further refined his concept of relevant technologies for rural areas, and began to wrestle with the prohibitive cost of creating workplaces in capital-intensive factories. Although his ideas and suggestions were largely ignored by Indian planners, for whom large-scale industrialization remained a preordained ideal, Schumacher persisted. Some of his ideas were borrowed: small-scale rural industries were already a standard development tool in China, for example. And he was deeply influenced by Gandhi who had coined a phrase Schumacher often quoted: 'Production by the masses, not mass production.'

Gandhi, who had popularized one of the first consciously 'appropriate' technologies, the spinning wheel, had said, 'The traditional old implements, the plough, the spinning wheel, have made our wisdom and our welfare ... India's salvation consists in unlearning what she has learnt in the last fifty years... You cannot build non-violence on factory civilization; but you can build it on self-contained villages.'[4]

But much of what Schumacher wrote was strikingly new, especially to a Western school of development thought grounded in classical economics and theories of aggregate demand. In 1964, he presented a paper at Cambridge entitled 'Industrialization through Intermediate Technology' which was widely published and republished, and at last he began to develop a following of people who saw the relevance of his message. Among them were David Astor, editor of the *Observer*, Lord Robens, Chairman of the National Coal Board, George McRobie, a fellow economist, and Julia Porter, Secretary of the African Development Trust. But Schumacher and his friends were uncertain as to how to proceed with an idea that seemed so *appropriate*, yet which received so little in the way of formal support. The governments of Burma and India had basically ignored the message. Britain's Ministry of Overseas Development was polite but uninterested, and the development establishment of the day was preoccupied with growth and Western models of large-scale industrialization – fuelled by apparently limitless supplies of cheap oil.

One day in May, 1965, a group of about twenty people joined with Schumacher in a meeting room at the Overseas Development Institute, and at last they decided to put their ideas into action. 'We had no money,' Schumacher recalled years later on a California radio programme. 'There were just a couple of friends of mine, like myself professional people with full-time work and families to support. But when you feel that something is necessary, you can't simply go on talking about it – you have to talk for a certain while, but then the moment comes – I tell you, a frightening moment – when you have to take the existential jump from talking to doing, even if you have no money.'[5]

In August, the *Observer* published a lengthy article that Schumacher had written months earlier, entitled 'How to Help Them Help Themselves'. It was as powerful and as succinct as anything he had ever written, and it came at precisely the right moment in the life of the nascent movement.

'Mass unemployment in developing countries,' he wrote,

> ... is being accepted as inevitable and unconquerable, even as 'necessary for sound growth', in much the same way as was the case in advanced countries before Keynes... Unemployment and under-employment in developing countries are most acute in the areas outside a few metropolitan cities; so there is mass migration into these cities in a desperate search for a livelihood: and the cities themselves, in spite of 'rapid economic growth', become infested with ever-growing multitudes of destitute people...
>
> No amount of brave statistics of national income growth can hide the fact that all too many developing countries are suffering from the twin disease of growing unem-

ployment and mushrooming metropolitan slums, which is placing their social and political fabric under intolerable strain.

The suspicion has been voiced (and cannot be dismissed out of hand) that foreign aid, as currently practised, may actually be intensifying this twin disease instead of mitigating it; that the heedless rush into modernization extinguishes old jobs faster than it can create new ones; and that all the apparent increases in national income are eaten up, or even more than eaten up, by the crushing economic burdens produced by excessive urban growth.

Rather than simply increasing aid flows, Schumacher argued that a change in direction was necessary, a change that would take aid agencies 'straight into battle' with the evils of poverty. This could be done by creating jobs in the areas where people live – the rural areas rather than cities. The workplaces would have to be cheap enough that developing countries could afford to create them in large numbers without unattainable levels of saving and imports. The employment should be in productive jobs requiring relatively simple production methods, producing goods for local consumption and relying to a large extent on local supplies of raw materials and finance. This could only be done, Schumacher said, through a conscious effort to develop an 'intermediate technology'.

Western technology has been devised primarily for the purpose of saving labour; it could hardly be appropriate for districts or regions troubled with a large labour surplus. Technology in Western countries has grown up over several generations along with a vast array of supporting services, like modern transport, accountancy, marketing and so forth: it could hardly be appropriate for districts or regions lacking these paraphernalia.

This technology, therefore, 'fits' only into those sectors which are already fairly modernized, and that means – some special cases apart – the metropolitan areas, comprising, say, 15 to 20 per cent of the whole population. What, then, is to become of the other 80 to 85 per cent? Simply to assume that the 'modern' sectors or localities will grow until they account for the whole is utterly unrealistic, because the 80 per cent cannot simply 'hold their breath' and wait: they will migrate in their millions and thereby create chaos, even in the 'modern' sectors.

The task, he argued, was to establish a tolerable basis of existence for the 80 per cent through an intermediate technology aimed at improving the productivity of traditional technologies, which were often in a state of stagnation or decay. This intermediate technology would be cheaper and simpler than the capital-intensive technology being imported from the West.

Some early critics argued that such a technology would waste resources, would produce less per unit of capital than known production methods, and could never be competitive. 'Dogmatic pronouncements are worthless,' Schumacher retorted. 'Let us have design studies and we shall see.'

(McRobie, p. 40) It would not be a long, expensive or particularly difficult task. Fragments of the technology that he foresaw already existed, both in the North and the South. What was needed was a systematic effort to collect them and develop them into practical blueprints for action.

> What stands in the way? Perhaps a kind of technological snobbishness which regards with disdain anything less than ultra-modern? Perhaps a certain callousness in the attitudes of privileged minorities towards the immense suffering of their homeless, jobless fellow-men? Or is it lack of imagination on the part of the planners in resplendent offices who find ratios and coefficients more significant than people?[6]

Public reaction was overwhelming and positive. Letters and support flowed in. In the following year the Schumacher group – which had been meeting in borrowed rooms – incorporated itself as a non-profit company and a registered charity. They called themselves the Intermediate Technology Development Group (ITDG), and their first donation – the fee from the *Observer* article – came from Schumacher himself.

Intermediate technology: a 'movement'

If, when ITDG started, there was a single basic criterion of what constituted 'appropriateness', it was low capital cost per workplace. There were likely to be four characteristics of a technology that would make it 'appropriate' to a developing society: it would be small in scale so that it could fit into small market situations. It would be simple, so that sophisticated manufacturing skills, organization and finance would be unnecessary. It would not be capital-intensive, and would therefore keep the cost per workplace down. And it would be non-violent. To Schumacher, the non-violence of a technology was an essential part of its appropriateness; it meant that an appropriate technology would be one that was completely under human control, that it would not have unintended side effects, that it would not cause social or environmental disruption.

Over the years since the basic formula – small, simple, cheap, non-violent – was developed, it has attracted a surprisingly large volume of academic and non-academic debate. Papers, theses and books have been written on the definition alone, arguing about finer points in the meaning of 'intermediate', and about definitions of appropriate: appropriate for whom? Appropriate to what? Who, after all, would be interested in an *inappropriate* technology?

Some observers insist on a clear distinction between 'appropriate technology', and 'intermediate technology'. The development of penicillin, for example, was very appropriate to the health needs it sought to address worldwide, but it was hardly an intermediate technology. Likewise the use of an earth satellite to predict a cyclonic disturbance is far from being an intermediate technology. But to villagers living in the low-lying delta areas of Bangladesh, the warning that this highly sophisticated, expensive technology can provide is more than appropriate. On the other hand, an intermediate technology, if it is going to be practical for development purposes, has to be an appropriate technology, and so sometimes the terms are synonymous.

In fact the example of communications technology in Bangladesh can be taken further. There, Grameen Bank, in a joint venture with Norwegian and Japanese firms, has equipped rural women with mobile telephones to rent out to their neighbours, giving them an income and their village a connection to the city and to the world. In the late 1990s a communications tycoon set up a mobile phone network in Mogadishu, helping to restore communications for business and hopefully for peace. A much less sophisticated form of communications technology was developed by British inventor Trevor Baylis in 1993 – a small radio powered by a crank-operated clockwork generator. By 1998 over two million of them had been produced in South Africa and distributed across the continent.

Intermediate technology is a relative term: a technology that stands somewhere between what is known – the traditional technology – and the modern. Thus in Africa, animal traction, falling half way between the hoe and the tractor, could be seen as an intermediate technology. In Asia, where animal traction is well developed, a power tiller might conceivably be the intermediate technology. But in some cases, neither of these examples might be appropriate. In Africa the most appropriate advance on traditional slash and burn agriculture might be better seeds, the use of fertilizer, an improved hoe. In Asia, the power tiller might not be the next logical progression on animal traction. It might be a better plough, improved irrigation, better organization.

In some cases, the definitions have overtaken common sense. George McRobie, who became Chairman of the Intermediate Technology Development Group when Schumacher died was, like Schumacher, far from dogmatic on definitions. Referring to 'small, simple, cheap and non-violent', he wrote that 'it is by no means certain that all four criteria can be satisfied in every case; *but any one of them, or combination of them, is of value for our purposes.*[7] Ten years later, however, the definition had become much more elaborate and insistent: 'Appropriate Technology is any productive process, any piece of equipment that meets *all* the following criteria (both emphases are in the originals):

- it meets the needs of the majority, not a small minority, of a community;
- it employs natural resources, capital and labour in proportion to their long-term sustainable availability;

- it is ownable, controllable, operable and maintainable within the community it serves;
- it enhances the skills and dignity of those employed by it;
- it is non-violent both to the environment and to people;
- it is socially, economically and environmentally sustainable.[8]

The irony in this definition – which combines dogmatism with an imprecision that begs for misunderstanding – is that virtually none of the practical ideas and technologies that Schumacher worked on during his lifetime would have survived it. Schumacher himself was impatient with 'those who get stuck on words; who start arguing with me that small is not *always* beautiful . . . these people who can't go beyond the words. This I consider an academic disease which is rampant.'[9]

In 1965, the concepts of appropriate and intermediate technology were simply that – concepts. Ten years later there were an estimated five hundred organizations, groups and institutions with an appropriate technology focus, and by 1980 the number had doubled to a thousand.[10] Some were small NGOs, others were attached to prestigious universities – St Gallen University in Switzerland, the Brace Research Institute in Montreal, and German universities in Oldenburg and Berlin. GATE was established in Bonn as the appropriate technology wing of the West German Foreign Ministry, and other institutions sprang up: TechnoServe in the US, the Groupe de Recherches sur les Technologies Appropriées (GRET) in France, and Appropriate Technology for Community and Environment (APACE) in Australia. Appropriate Technology International (ATI) in the United States had an unusual start: inspired by Schumacher's ideals, and by his widely-read book, it was funded by USAID but was given a direct mandate from Congress as a private, non-profit organization. Although some organizations were oriented towards international development assistance, others such as the National Center for Appropriate Technology, founded in Montana in 1976, and ATTRA (Appropriate Technology Transfer for Rural Areas), which started in Tennessee in 1987, had a largely domestic focus.

Far from being centred in the North, however, the idea quickly took root throughout the South. With advice from ITDG, the University of Science and Technology in Kumasi, Ghana established a Technology Consultancy Centre in 1971. The University of Zimbabwe founded a similar institution, again with advice from ITDG, in the early 1980s. Committees, groups and organizations were established in Zambia, Kenya and Tanzania. Two institutions were opened in Botswana – the Rural Industries Innovation Centre at Kanye and the Botswana Technology Centre in Gaberone. There were dozens of others: the South Pacific Appropriate Technology Foundation, the Centro Co-operativo Tecnico Industrial in Honduras, the Centro Meso

Americano de Estudios Sobre Tecnologia in Guatemala, and many in India, such as the Appropriate Technology Development Association in Lucknow.

Co-ordinating bodies were established: Volunteers in Technical Assistance (VITA), which predated ITDG by six years, had been established to disseminate knowledge about known, relevant technologies to developing countries. SATIS (Socially Appropriate Technology International Information Services) was founded in 1982 to disseminate information about technology produced by other organizations. And the Manila-based APPROTECH, supported initially by ATI, was formed as an Asian regional information centre on appropriate technology.

Everywhere there were experiments with windmills, solar energy, latrines and energy-efficient stoves. Experimental pumps of every sort began to dot the yards behind workshops – pumps made of bamboo, wood, plastic; pumps powered by the sun, the wind, bicycles and oars. Small biogas plants producing methane from animal waste, a technology developed concurrently in India and China, began to appear everywhere.

It was in some of these initial enthusiastic projects that the first weaknesses in the AT movement began to appear. The movement had been formed of an uncritical coalition between two distinct groupings. The first, predominantly concerned with the economic and social development of the Third World, was rooted mainly in the non-governmental and academic communities. The second was largely a consumer-based movement made up of people disenchanted with mainstream industrial and technological development in the North. Growing out of the counter-culture of the late 1960s and early 1970s, this part of the movement focused less on the technology and objectives of organizations like ITDG and more on an 'alternative' technology aimed at simpler, less wasteful Western lifestyles. The Jade Mountain Access Company in Boulder, Colorado, for example, began life as a Northern California natural food store called 'Evergreen', backing into appropriate technology almost by accident. By 1999, the company's website was offering to 'assist those who are striving to live appropriately on the planet by choosing tools accurate to a situation, tools that do the job and no more . . . Today our catalogs describe tools, ideas and energies that fit E.F. Schumacher's code for problem-solving in today's world: small, simple, inexpensive and non-violent.'[11]

That Schumacher's message spoke so eloquently to both streams is a mark of its universality. But the Western counter-culture, searching for a return to simpler times, often had little to say to the poor of the Third World, armed with rudimentary tools, suffering from ill health in debilitating climates, eking poor livings out of poor soil or earning pennies a day in crowded factories. The Walden Pond school of appropriate technology, which carried over into the environmental movement, forgets – or never knew – that Henry David Thoreau did not brave the harsh frontier. He may have lived a bucolic life in the Massachusetts woods, but even in 1845 his woods were not far from civilization, and the pond was not far from his mother's house where he often had dinner when his own larder was bare. This side of AT is a protest, often a very legitimate one. But it is neither intermediate, nor is it particularly relevant to the parched maize fields of the Sahel, the harsh Altiplano of Bolivia or the crowded slums of Manila.

This is not to diminish the importance of Thoreau nor, more particularly, of the Northern environmental movement. Much of it, however, has been considerably more concerned with trees, birds and wildlife than with people. This may be natural enough in the North where most people have most material needs taken care of, and where industry and consumerism have paved over a sea of Walden Ponds, converting them into parking lots and hamburger stands. It is far from appropriate in the South, however, where much environmental degradation is caused by poverty rather than the reverse. That is why tree-planting programmes in the South often have limited success until the poverty of people who survive by cutting down trees is taken into consideration. This theme will recur in subsequent chapters which deal with energy and a growing convergence of interest between the development and environmental movements. It will suffice to note here that the idea of 'transferring' capital-intensive technology willy-nilly to developing countries is no more arrogant than attempting to transfer Western environmental worries without adaptation, and without careful reference to the very different circumstances in which Southern people live.

Luddites in reverse gear: the critics

The 'AT movement', sometimes advanced and occasionally retarded by its informal alliance with the Alternatives Movement, went through two distinct phases in the 1960s and 1970s. The first was a kind of 'messianic phase' during which it was assumed, partly because of a lack of funding, that once the problem was understood, solutions would fall easily into place. It reflected Schumacher's early idea that the task would not be 'long, expensive or particularly difficult'. This proved to be far from true, but in the early years it was easier to indulge in criticism of the prevailing development ethic and its emphasis on Big, than to find cheap, workable technologies that could demonstrate the effectiveness of the alternative.

As the 1960s – the 'First Development Decade' – drew to a close and the failings of the basic growth model started to become evident, however, more funding became available for alternative approaches and for NGOs. Schumacher, no longer out in the cold, now saw that in itself, the message was not enough. In Britain, ITDG began to form groups or panels of engineers and designers, builders, agronomists, and health and energy experts who could actually begin to work on testing and adapting ideas. In some cases what they came up with were existing technologies such as small-scale textile equipment or corn grinding machines, well tested but unknown to potential users. In other cases, such as brick making or soap production, they took traditional technologies and improved them. A third approach was the simplification of a modern technology such as hydro-electric power or the manufacture of nuts and bolts. And in some cases there were completely new technologies such as solar energy.

This second phase of the AT movement, which lasted well into the 1980s, is regarded by some as the 'arrogant phase', a period dominated by technology and hardware – what one AT worker called 'the better mousetrap phase'. It was, to a large extent, characterized by the same arrogance that fuelled mainstream aid agencies, and its failures can be seen rusting behind the workshops of dozens of AT institutions around the world. Part of the problem was a fixation with technology, at the expense of dissemination. Dissemination, it was assumed, would follow naturally enough if an idea was a good one. Alternatively, diffusion was the responsibility of others – government, NGOs or perhaps the private sector. And poor diffusion was a reflection on them rather than on the technology.

Biogas is perhaps the archetypal example. Tens of thousands of successful methane gas digesters were already in operation in India and China when the AT movement began to work on the technology. Through the 1970s, literally thousands of prototypes were developed, redeveloped and field tested in other Asian countries, in Latin America, the Caribbean and all across Africa. The result has been mixed. In Northern Pakistan a local NGO, pressed by donors to introduce the technology into the villages with which it worked, discovered that gas production decreases as temperatures drop. When the gas is most needed for heating and cooking, it simply isn't available. In other countries, it was found that the number of animals required to fuel a digester made it inaccessible to the poorest families. In addition, cows had to be kept in a restricted area so the dung could be collected properly. Attempts to organize several families around one gas plant were often futile – in part because not enough gas was produced for the number of people involved. In some parts of the South it was as difficult to get people excited about working with animal faeces as it would be in Pittsburgh – a problem that did not exist in China and India where animal dung had long been a traditional source of fuel.

Experiments in Botswana illustrate the problem. Botswana has nothing, if not cattle. But after eight years of expensive experimentation at the Rural Industries Innovation Centre, only 12 biogas plants had been put into field situations. Of those that remained in operation after two years, most were experiencing severe operational and management difficulties.[12]

The charge that diffusion was given a back seat during this phase of the AT movement's development is not universally true, but it is valid to say that widespread diffusion of technologies that seemed genuinely relevant proved far more difficult than had been imagined. In retrospect, this is hardly surprising. The difficulties, however, fuelled a reaction in the official development and academic communities. Quick to find fault, especially fault that allowed a guiltless return to mainstream economics and the continued export of high-cost Northern goods and services, the attack – which continues – was mounted on technical, economic, cultural, social and intellectual grounds. It was, in a sense, a reverse Luddism. Ironically, these modern Luddites no longer attack modern, job-eliminating technology; they take their intellectual cudgels to innovations which seek to do the opposite.

It is easy enough to find examples of how a particular attempt to develop or disseminate a technology failed. Much of the early AT effort revolved around hardware, developed by engineers who loved tinkering and invention. Economic, social, organizational and institutional concerns were often not their long suit. These issues, correctly enough, were left to others. Many of the AT organizations with which they worked, however, were typical of well-funded NGOs – staffed by generalists, careless about evaluation, with levels of income that bore more relationship to the quality of their fundraising literature than to any actual success or failure taking place five thousand miles or more away from the support base.

Some critics began to debunk the notion of 'technology choice', arguing that there is usually only one way of efficiently and economically pursuing an activity, only one readily available choice that makes sense in a given situation.[13] This ignores one of the underlying purposes of intermediate technology, however, which is to rethink and redefine what 'makes sense'. It is also argued, as the McNamara quotation at the start of this chapter does, that the AT movement, stuck in unrealistic theories of redistribution, is somehow opposed to growth. It is perhaps worth noting that the concept of redistribution with growth, widely accepted as gospel in the 1970s and popularized in a book of the same name, emerged in part from the World Bank's own Development Research Centre.[14]

Reflecting on the lessons of technology in history, Schumacher argued that technology is both determined by, and is a result of, factors within a society; it is not only, or always, or even very often something that developing countries can order cold from a catalogue with hopes of instant success. Schumacher criticized the pathological inability of mainstream economists to think beyond *quantitative* growth. 'Instead of insisting on *the primacy of qualitative distinctions*, they simply substitute non-growth for growth, that is to say, one emptiness for another.'[15]

Schumacher was equally critical of Adam Smith's 'invisible hand' concept of the market economy as a tool for guiding the choice of technology in developing countries. The influence of donors, governments and individuals on the choice of technology can be enormous, as well as enormously destructive. This imposes another kind of invisibility on the choice of technology in developing countries, one that is not well represented by precedent in other times and other places. Whether or not the 'invisible hand' of the market-place is a useful idea, where technology choice is concerned, the issue is as much one of invisible attitudes and policies as one of a fettered or unfettered marketplace.

Richard Eckaus, an American economist, was typical of many critics.[16] Most of the barbs were aimed at the movement's fringe elements, at its early beginnings, and at the more generic weaknesses of NGOs. Eckaus argued, for example, that appropriate technology is a nebulous concept without clear definition, one unsupported by economic, social or scholarly analysis. Like other critics, he clung to 'better mousetrap' times, claiming that the movement was technologically determinist, a hardware store in search of customers. In fact by the mid-1980s, most of the best AT organizations had moved away from the mousetrap phase towards what ITDG called an 'enabling phase', one which sought to address the *causes* of limited uptake.[17] In some cases this had to do with the technology itself. In others it was the 'invisibility' of attitudes and policies framed in ivory towers, corporate boardrooms and donor headquarters. In others it was a matter of organization. What ITDG, ATI and other institutions had discovered was that techniques of organization were as relevant to the transfer of technology as the technology itself. This was especially relevant to poverty alleviation; for example, in the organization of landless people in Bangladesh around a known technology – irrigation pumps, or in the vertical integration of a well-known but once moribund technology, silk spinning and weaving.

A new AT phase can be perceived in the 1990s, one that at first glance might be termed a 'madly off in all directions' phase. For some – like the Dutch organization TOOL, faced with an extinction of government support in 1998 – it meant closure. For others, like TechnoServe and ATI in the United States, undoubtedly influenced by funding trends in USAID, it meant a dramatic shift from their technology roots towards small business development, in which technology seemed to play a backseat role to the commercial context and opportunity into which it was intended to fit. Many organizations became concerned with environmental issues. For some of the engineers who grew up with the movement, it was as though appropriate technology, taken over by wannabe businessmen and social scientists, had lost its way. In

ITDG, process, participation and sustainability took on more prominent meaning in the 1990s, with as much emphasis placed on the potential user of a technology as on the technology itself. 'Participation' – emphasis on the user – while perhaps new to some appropriate technologists, and even to parts of the broader aid enterprise, is not new to the market-place. A product's price, utility and the general acceptability of its shape, size, colour, taste or smell will all affect its uptake by the intended consumer. In the case of AT, the 'product' is new technologies and processes, and the 'consumer' is the intended user. Ignorance of what the user/consumer knows, understands, wants *and is capable of using* has a long history of affecting sales in a negative manner. Recognizing this, ISO 9000, a series of rigorous international standards for quality management systems, places heavy emphasis on client participation in the design, testing and refinement of a new commercial technique, tool or product. Before it will be certified as ISO 9000 compliant, a company must demonstrate client involvement.

In reality, while changes among AT organizations may appear dramatic or overly business-like, and while there may be claims with each passing decade of greater sophistication and better congruence with development concepts (and buzz-words) of the day, many of the changes are variations on themes that were there from the start – better understood perhaps, more carefully nuanced to meet local needs, but not greatly different from the basic ideas that Schumacher put forward in the 1960s. A critic of many of capitalism's flaws and failings, Schumacher was nevertheless very clear about small business, saying that 'in small-scale enterprise, private ownership is natural, fruitful and just.'[18] Where participation is concerned, he subtitled *Small is Beautiful*, 'A Study of Economics as if People mattered.' He believed that an intermediate technology, 'a technology with a human face, is in fact possible; that it is viable; that it reintegrates the human being, with his skilful hands and creative brain, into the productive process.'[19]

Small revisited

Since *Small is Beautiful* was published, there have been enormous changes in technology, in raw materials, in production techniques, information systems and product development. In many ways, the gaps Schumacher sought to fill have become wider. But much has happened to revalidate his vision, and the messages contained in *Small is Beautiful*. The first is that the appropriate technology concept has matured, and its proponents have become more professional. Although some of the weaknesses of the 1970s and 1980s persist, there is a growing body of success stories and much greater sophistication in dealing with the societal and political obstacles that in the past have limited progress.

The second is a growing recognition in the West as a whole that Big is not, in itself, better. Mass production and giant industries grew out of a particular approach to organization and out of ideas about the primacy of price and the importance of economies of scale. There have always been dissenting views, however. In the 1930s and 1940s, for example, Joseph Schumpeter argued that it is the firm, not the market, that determines success in industry. It is not so much price competition, he said, but 'competition from the new commodity, the new technology, the new source of supply, the new type of organization,' that determines the wealth and progress of a nation.[20] It is the entrepreneurial firm, which seeks strategic innovation in product, process and organization, and through constant improvement, that has the ultimate advantage over hierarchical firms rooted in single-minded price competition.

This lesson, forgotten or ignored in Europe and North America, was rediscovered and used to excellent advantage by Japanese firms in the 1970s. The large body of literature available on the reasons for their success confirms Schumpeter's views. But there are other reasons for the Japanese success, one of the most prominent of which lies in the relevance of Small. Price was important, but it was not the only factor, nor was it necessarily the most important factor. The most successful Japanese firms, which by world standards are enormous, have focused clearly on the customer – customer preference, customer need, customer taste. Within the best firms, there is an emphasis on innovation, and on close, personalized coordination between divisions, and between managers and the shop floor.

An automobile has approximately 20 000 parts. According to Michael Best, the average Japanese car-maker produces only 30 per cent of the parts itself, subcontracting the rest to smaller firms. The average American car-maker produces about half of its own requirements, and thus operates much larger and more complex plants. But the story doesn't stop there. General Motors deals with 3500 subcontractors, whereas Toyota deals with only 300. The difference between the complexity of the American network and the Japanese is as day is to night. But the major difference is that Toyota's subcontractors, with whom it is able to develop a close, personalized relationship, subcontract much of their own work to a second tier of 5000 firms, which in turn deal with a further 20 000 smaller third- and fourth-tier firms.[21]

Within the American firms examined by Peters and Waterman for their best-seller, *In Search of Excellence*, the same principle was at work: the creation of what Western economists have traditionally disdained as 'suboptimal' units, along with decentralization, encouragement of independent thinking, innovation and new approaches to old problems. 'The message from the excellent companies we reviewed was invariably the same,' they wrote. 'Small *is* beautiful.'[22]

The phenomenon is not unique to Japan. In the US, between 1976 and 1986, the number of new businesses nearly doubled, while GNP grew by only a third. The average contribution to GNP per firm dropped from $245 000 in 1980 to $210 000 by 1987, falling in every industry except retailing and farming, thus putting the lie to the common assertion that growth in small-scale enterprise is simply a shift from manufacturing to service industries. In Britain, between 1982 and 1984, firms employing fewer than 20 people created an estimated one million new jobs, offsetting 750 000 job losses in large firms.[23] Between 1975 and 1990, the top 500 American companies added no new jobs to the economy whatsoever,[24] and after that they seemed to specialize in laying off as many workers as possible.

Perhaps the most striking example of new industrial thinking can be found around the 'industrial clustering' phenomenon, first and most prominently found in the 'Third Italy', a region between the industrial heartland of the Italian north and the agricultural regions of the south.[25] Characterized by a vast number of small firms, the Third Italy is not unlike nineteenth century Birmingham, or the Connecticut Valley which fostered the Springfield Armory and modern American production methods. Of Italy's twenty regions, Emilia-Romana was the fastest growing through the 1970s and 1980s. Its 350 000 registered firms had an average of only five employees each, and an estimated one third of the work force was self-employed. And yet Emilia-Romana had the highest per capita income in Italy and accounted for 10 per cent of Italy's exports. With imports to the region accounting for only 4 per cent of the country's total, Emilia-Romana contributed an international trade surplus of $5 billion to the Italian economy every year. Similarly, in the 'tile valley' of Sassuolo, tile production increased 186 per cent between 1973 and 1997, and the percentage of exports rose from 30 per cent of the total to 70 per cent.[26] Part of the Italian success story is based on the country's machine tool industry, the fourth largest in the world after Germany, Japan and the United States, with sales of $3.45 billion in 1997. Here too, however, the firms are small by international standards, with an average of 70 employees in 450 firms, compared with 200 in Germany's 320 firms.[27]

Similar examples of clustering can be found in the South. The Pakistani city of Sialkot is home to an impressive industry in stainless steel surgical instruments. In the mid-1990s, some 350 firms and 2000 ancillary service providers were producing half of the world's demand for low-quality clinical instruments, and 10 to 15 per cent of the demand for high-quality surgical instruments. More than 70 per cent of the producers were small firms with fewer than 20 employees.[28] The city of Ludhiana produces 95 per cent of India's woollen hosiery, 85 per cent of its sewing machine parts, and 60 per cent of its bicycles and bicycle parts. While there are some large firms in Ludhiana, small firms are the core of its output. More than half of Mexico's leather and footwear industry is located in the vicinity of Leon, and nearly a quarter in Guadalajara, much of it in small and medium-size firms.[29] Clustering, of course, is not exactly new. Areas like London's Fleet Street (newspapers), Paternoster Row (books) and Bread Lane recall a long-standing tradition of clustering.

Unlike the high-tech silicon valleys, glens and parks celebrated in the business sections of European and American newspapers, today's small Italian, Pakistani, Indian and Mexican workshops are largely devoted to lower-technology industries in leather, textiles, metalworking and automobile parts. Their success lies not only in the smallness of the firms and their clustering, but in their organization, in their productive associations and consortia, and in the physical, monetary and policy support they receive from government. The lessons of Italian, Indian, Pakistani and Mexican clustering are important for small enterprises in other developing countries, where the shadows of similar phenomena can often be discerned.

A third factor that revalidates Schumacher's message is a growing awareness of the informal sector. Unorganized, disorganized, clandestine and usually illegal, the informal sector has been neglected by aid agencies, denigrated by economists and harassed by officialdom. And yet studies are revealing how widespread and how important it is. In most Southern cities, it represents as much as half the workforce; in Bombay it occupies a full 55 per cent of the labour force. In Peru it covers 47 per cent of the population on a permanent basis, accounts for 61 per cent of the person-hours worked, and contributes 38 per cent to the gross domestic product.[30] In Karachi, less than one third of the demand for housing is being met by the formal sector, and 72 per cent of the city's transportation is provided by 'private sector' mini buses, most operating in a semi-illegal capacity and purchased entirely through informal-sector loan arrangements. According to the Karachi Master Plan 2000, 76 per cent of the city's turn-of-the-century workforce is employed in the informal sector, up from 66 per cent in 1972.[31]

The disaster looming behind these figures is compounded by other statistics which show that the formal sector is not creating jobs at a rate anywhere near the demand, and that the jobs it does create are as prohibitively expensive as they were when Schumacher wrote on the subject in the 1960s. Zimbabwe's 1986–90 National Development Plan, for example, forecast an increase in wage employment of 21 000 workers, with an expected investment of Z$1.4 billion. This put an amazing price tag of US$26 000 on each job.[32] Estimates of the number of jobs required annually in order to 'mop up the backlog' and to deal with new school leavers in Zimbabwe ranged from a low of 160 000 to more than a quarter of a million.[33] At this rate, using the lower figure, it would have cost over US$4 billion per annum to create the necessary number of jobs, an amount equivalent to 90 per cent of the country's entire GNP. Meanwhile, as a result of a structural adjustment programme introduced in 1991, Zimbabwe actually looked forward to job *losses* of 20 000 in industry, 10 000 in government, and a further 2000 in parastatals.[34]

The same numbers were playing themselves out in Bosnia a few years later. In the first two years after the 1995 Dayton Peace Accord, donors committed an exceptional $3 billion to a medium-term Priority Reconstruction and Recovery Program. With massive unemployment and 1997 GDP at less than half its 1990 level, some donors opened lines of credit for medium and large-scale enterprises. USAID's Business Development Program had made 123 loans by the end of 1997, totalling about $70 million, and resulting in an estimated 10 600 jobs.[35] At $6 600 per job, this looked better than in Zimbabwe – but no more realistic, with the cost of generating full employment at that rate totalling some $8 billion.

There are only two solutions, if people are to work. The first is that the cost per workplace in the formal sector will have to come down dramatically, as Schumacher said it must if chaos is to be reduced or avoided. The second is that the informal sector will have to expand to make room for large numbers of new entrants. This cannot be done if governments continue to ignore it, hound it and hide it behind tin-sheet fences. Donor agencies have begun to realize its importance, and there is a growing number of indigenous lending institutions, especially in Asia – Grameen Bank, SEWA, BRAC and BKK, for example – that have developed successful programmes for lending to people without assets. But lending alone will not formalize the informal sector. Nor will a few hundred takas or rupees or pesos make a poor person financially self-reliant. The policy framework in which a credit programme operates is critical to the productivity of any investment, whether it is a hundred taka or a million. And genuine investment opportunities for poor people, opportunities that go beyond the obvious – poultry, petty trade, handicrafts – require the kind of research and development with which the appropriate technology movement has been engaged for more than thirty years. In a sense, greater integration of the informal sector requires a kind of reconstruction of the economy and of economic thinking. Much of Schumacher's writing was concerned with 'reconstruction', flowing, no doubt, from his experience between 1946 and 1950 as economic adviser to the British Control Commission in Germany.

There is a final phenomenon which revalidates Schumacher's philosophy. That is the overwhelming evidence that the continuing pre-eminence given to indiscriminate growth strategies has done little or nothing to reduce poverty. There are more poor people today than there have ever been, and the numbers show little sign of diminishing. Schumacher's predictions about growing misery, urban poverty and social unrest have all come true with a vengeance that wreaks horror on the lives of increasing millions.

'Our scientists,' Schumacher wrote,

> tell us with the utmost assurance that everything around us has evolved by small mutations sieved out through natural selection . . . even the Almighty is not credited with having been able to create anything complex. Every complexity, we are told, is the result of evolution. Yet our development planners seem to think they can do better than the Almighty, that they can create the most complex things at one throw by a process called planning, letting Athene spring, not out of the head of Zeus, but out of nothingness, fully armed, resplendent, and viable.[36]

It is perhaps worth returning to Schumacher's comment that today's poverty is 'a monstrous, scandalous thing which is altogether abnormal in the history of mankind.' In the last pages of his final book, *A Guide for the Perplexed*, published shortly after his death, Schumacher recalled the *Divine Comedy*. In it, Dante wakes up to find himself in a horrible dark wood from which he cannot escape without first descending into the Inferno, in order to appreciate the reality of sinfulness. 'The art of living,' Schumacher wrote, is always to make a good thing out of a bad thing. Only if we *know* that we have actually descended into infernal regions where nothing awaits us 'but the cold death of society and the extinguishing of all civilized relations', can we summon the courage and imagination needed for a 'turning around'.

> Can we rely on it that a 'turning around' will be accomplished by enough people quickly enough to save the modern world? This question is often asked, but whatever the answer is, it will mislead. The answer 'Yes' would lead to complacency; the answer 'No' to despair. It is desirable to leave these perplexities behind us and get down to work.[37]

CHAPTER VII
Farmers, food and forests

Glendower: 'I can call the spirits from the vasty deep.'
Hotspur: Why, so can I, or so can any man. But will they come when you do call for them?'

William Shakespeare, *Henry IV, part 1*

In the search for new approaches to rural development in the 1960s and 1970s, it was sometimes forgotten that most innovations in farming have been made by farmers, not by research scientists. Like factory workers and mechanics, farmers do not write much history, and therefore tend to be forgotten, especially when they are poor and speak a strange tongue. It was 'primitive' farmers, however, five and even ten thousand years ago, who domesticated almost all the crops and vegetables in use today, and who domesticated animals for food and for draught power. As any good extension officer knows, the innovative spirit is alive and well in almost any farming community.

Much of what farmers do is devoted to food production. With the world's population likely to grow by three billion people in the first quarter of the twenty-first century, food production has become an increasing concern. Almost three-quarters of the people in low-income countries today earn all or most of their income and subsistence from agriculture and related activities. Agriculture, therefore, faces a double challenge: it must produce food at a rate of expansion never before known. And it must provide most of the world's population, including the increase, with a livelihood.

Return to Oz: the green revolution

During the first half of the twentieth century, colonial agricultural research stations were established in Africa and Asia to work on the development of hybrids and higher-yielding varieties, mainly of cash crops. The 1950s and 1960s, however, were a period of impressive breakthroughs in food crop research. The first was an adaptable dwarf wheat, developed in Mexico by

Norman Borlaug under the auspices of the Rockefeller Foundation. Because most of the world's poor live on a rice diet, Rockefeller, joined by the Ford Foundation, turned towards Asia, and in 1960 founded the International Rice Research Institute in the Philippines. Gathering some 10 000 strains of rice from around the world, scientists began the task of cross-breeding, and within a relatively short period – combining a Taiwanese variety with one from Indonesia – they developed what became known as IR-8, a rice capable of doubling traditional yields.

National agricultural institutions further developed the new varieties to suit local conditions. The impact was enormous. In 1964, an estimated 200 acres were devoted to high-yielding cereal production throughout Asia. By 1973, 80 million acres in Asia and North Africa had come under the new varieties, and by 1991, 74 per cent of all Asian wet-season rice consisted of modern varieties.[1] Pakistan was transformed from a grain importer into a net exporter in only seven years; Indian wheat production increased from 11 million tons in 1965 to 26 million tons in 1972, a rate of growth unmatched in history. Between 1982 and 1992, the production of cereals grew in Asia by 25 per cent, and in Africa by 41 per cent.[2]

The shortcomings of the green revolution have perhaps received more attention than its achievements. Genetic engineering, globalization of the food system, and the extension of industrial patenting into the seed business have contributed to a rapid and serious decline in genetic biodiversity. Critics have been justified in their worry about the narrow genetic base of some plants, and their stability against weeds, disease and pests. Because new varieties are more dependent on fertilizer and pesticides than traditional agriculture, investment costs are often beyond the means of small farmers. Likewise, the organization of improved irrigation favours larger farms, while government distribution and credit schemes favour people with collateral. Many are unable to make the investment, but many others who might be able to do so shy away. The poor are understandably 'risk-averse': a single crop failure can mean the loss of everything – land, livelihood, family, even life itself.

Nevertheless, there is now a large body of research on how high-yield varieties (HYVs) have affected the poor, and the message is not all gloomy.[3] Poor farmers *have* adopted the HYVs, and because the new varieties are more labour intensive, there has been a consequent increase in employment. Food prices, which account for the bulk of a poor family's spending – in rural and urban areas – have been reduced as well. The impact of the new varieties on the farming practices of the marginal farmer has, however, been limited. The new strains were developed for average or better than average land, not the remote, infertile patches on which the poorest inevitably live. The increases in food production of the 1960s and 1970s, therefore, mask worrisome statistics: the bulk of the world's poorest people are functionally landless, or live on small plots of poor land, overusing the resource base and eking out less than subsistence lives with traditional, low-productivity techniques. As important as this problem is the fact that there has been a marked slowdown in food production since the tremendous advances of the 1960s and 1970s. Although 1996 marked the largest global grain harvest on record, the biggest gains of the green revolution have already been made in many countries, and in some, food production is barely keeping pace with population growth. In Africa it is not keeping pace. There it is estimated that the number of undernourished people could rise to an estimated 300 million by 2010, with a per capita food supply of only 2170 calories a day (less than the intake of those defined as being among the 'absolute poor').[4]

Deforestation: cutting down tomorrow

Tropical deforestation is a relatively new environmental issue, publicized in part by the 1987 Brundtland Commission. The entire Brandt Report, issued six years earlier, devoted only three paragraphs to the subject. Until about 1990, the most authoritative estimates on tropical deforestation were based on a 1980 FAO study which calculated the annual loss at 11.4 million hectares.[5] More recent studies indicate that the situation is much worse. At 16 million hectares each year, the loss represents an area five times the size of Belgium.[6]

One of the major causes of deforestation is poverty – the poverty of individuals and the poverty of nations. This is not a question of 'blaming the poor', nor of absolving the rapacity of people in the North and the South who cut down trees for profit. It is a simple statement of fact. The world's poorest people generally live in the most environmentally vulnerable areas. In Africa, 51 per cent of the poorest people live on marginal land that is highly susceptible to soil degradation and erosion, floods and other natural calamities. In Asia the figure is 60 per cent, and in Latin America it is an incredible 80 per cent.[7] The most prominent cause of deforestation is related to population pressure, the need these people have for greater security, and the conversion of forest land to agriculture. The second contributory cause, logging, does not have to lead to deforestation if it is properly managed. In most poor countries, however, it is not managed at all. The third is a growing demand for pulp, fuelwood, fodder and other products unavailable at affordable prices except from the forest.

The demand for 'affordable prices' can be both a cause and product of corruption, and of laws without teeth. The illegal export of timber, for example, is rampant in many developing countries, representing as much as $65 million in trade each year in Ghana and $241 million annually in Papua

New Guinea during the 1990s. An estimated 80 per cent of timber extraction in the Amazon is done illegally, and in Cambodia, the Khmer Rouge financed itself to the tune of $10–12 million each month in 1995 and 1996 by selling stolen timber to Thai logging firms.[8]

The effects of deforestation are widespread. Exposing weak soil to erosion, deforestation leads to a loss of biological diversity, the creation of permanent wasteland, flooding and the siltation of reservoirs. And the game, birds, berries, nuts, resins, spices, honey, oil and other products associated with the forest disappear. In Indonesia, such products were worth $125 million in annual exports in the early 1980s.[9] Deforestation also contributes to the global carbon cycle. Second only to the burning of fossil fuels in the production of atmospheric carbon dioxide, deforestation is directly related to the greenhouse effect and global warming. Deforestation and land-use change in the tropics, in fact, account for 33 per cent of annual human-related carbon dioxide emissions. Brazilian forest cover was lost at an annual average rate of about 19 000 square kilometres annually between 1976 and 1996, much of it to deliberately set forest fires that released hundreds of thousands of tons of carbon into the atmosphere every year.[10] In Indonesia, fires started by pulp, palm oil and rubber plantation owners to clear land blanketed the entire region in smoke and haze for months during 1997, destroying at least 2 million hectares of forest.[11]

Sustainability: the myth

'Sustainability' is one of many good words that development agencies have hijacked, using it to justify and disguise so many things that its meaning has become lost in a swamp of jargon. The dictionary defines 'sustain' in several ways: to withstand; to maintain or prolong; to be supported from below; to provide or give support to – as in supplying necessities; to keep up vitality or courage; to affirm the justice or validity of an endeavour. But the word has come to mean other things in the development lexicon. It has become a euphemism, subjectively used, for virtually everything 'good' and benign. But when hard project-funding decisions must be made, it generally comes to rest on the notion of *self*-sustainability – the capacity of a project or endeavour to stand on its own feet financially when external support is withdrawn. This fundamental alteration can lead to programming efforts that are driven by narrow, short-term economic concerns, and which fail to appreciate the broader context within which genuine sustainability must exist.

Genuinely sustainable agriculture is directly related to the sustainability of resources. Renewable local resources, such as land, air and water, are sustainable only as long as they are used properly. Erosion, upland deforestation, chemical pollution or incorrect water usage can all have detrimental effects on them. Non-local resources such as seeds or fertilizer are another factor. If adequate and timely supplies can be maintained at stable prices, there is no reason to assume that these detract from sustainability. Changes in national or international prices, however, can have a major impact on sustainability. Structural adjustment programmes, for example, which emphasize agricultural subsidy reductions, may damage sustainability, especially at the margins where farmers and consumers are less able to withstand price increases. Removing a subsidy on fertilizer in Sri Lanka, for example, had a direct negative impact on both food producers and consumers.

Food production cannot be divorced from imports or, therefore, from the exports needed to pay for them. Even if they could do without imported fertilizer and pesticides, countries need roads and trains and trucks in order to distribute seeds and fertilizer, and to move agricultural produce to market. They need fuel, bitumen for roads and steel for rails and agricultural implements. All of these things reflect wage rates and inflation in industrialized countries. All reflect the price of oil, which moved from $3.22 a barrel in 1973, to variations on a theme ranging between $15 and $50 a barrel (at constant prices) over the next two and a half decades.

The idea, therefore, of perfect national food security, of a country which requires no imports to sustain its own food production, belongs to another time and another place. In other words, countries must export. Yet since 1950, the prices of what most poor countries export have declined in both real and relative terms, falling to record lows in the mid-1980s and continuing to decline through the early 1990s. This resulted in a greater emphasis on export crops, and on usually futile attempts to do more than tread water. In 1960, for example, a ton of bananas could buy 13 tons of oil. By 1972, it paid for only 1.6 tons. In 1960, a ton of sugar would purchase 6.3 tons of oil. By 1982, the same ton of sugar would purchase less than three-quarters of a ton. Thailand increased its exports of rubber by 31 per cent between 1984 and 1985, yet because prices changed, it earned eight per cent *less* in rubber revenues, a drop from $262 million to $242 million. Between 1984 and 1994, the price of palm oil fell 55 per cent in constant dollars, cocoa fell 64 per cent and tea fell 67 per cent. And the price of coffee dropped 35 per cent during the same period.[12] A surge in the prices of some commodities in the mid-1990s appeared to offer hope. Coffee prices, vegetable oil, rubber, cocoa and cotton were all up. By the end of the century, however, the rollercoaster had taken another dive. Cocoa dropped 27 per cent between 1997 and 1999, arabica coffee fell 47 per cent, palm oil was down 17 per cent and cotton had fallen 31 per cent.[13] Price fluctuations of this nature have a devastating effect on farmers, and on the economies of the countries that depend on them.

Increased emphasis on cash crops can reduce the amount of land available for food production, but it has other, sometimes more far-reaching effects. The government of Indonesia, for example, encouraged an increase in cassava exports by doubling the farmers' price in 1985 and again in 1987. The result was that many Javanese farmers switched from mixed farming to cassava, removing terraces and water conservation structures that had been built at great cost and which in the past had prevented erosion.[14]

As a concept, therefore, agricultural sustainability requires careful thought, especially by donor agencies. It is far more than an investment in a self-financing enterprise. There are complex trade-offs that must be considered: trade-offs between export crops and food production; trade-offs between productivity and equity; trade-offs between short- and long-term production; trade-offs between investment and productivity; trade-offs between genuine agricultural sustainability or food security and the prevailing system of international trade.

Agroforestry

This is not to suggest that nothing can be done. On the contrary, as the limitations of the green revolution became more apparent in the 1970s, greater attention was focused on technologies and approaches that could reduce the cost of food production and increase the productivity of people living on marginal lands. In India, for example, huge investments have been made in the construction of earthworks to reduce farm soil erosion. Simply to protect the existing agricultural land base from further erosion, however, this strategy could cost as much as $14 billion. An alternative, already well established in Fiji and hilly areas of some Caribbean islands, was under trial in the late 1990s, however. It is the simple use of a coarse perennial grass, used in conjunction with contour planting and hedges. The new method is between one-tenth and one one-hundredth the cost of earthworks.[15]

Agroforestry, known for centuries in some countries, began to gain widespread acceptance in the 1980s. Basically, agroforestry is the combination of crops with trees and shrubs that provide nutrients to the soil through their root systems, through the 'green manure' they produce, and their mulch. For centuries, Javanese garden culture has practised an intensive form of agroforestry, combining horticulture, animal husbandry and tree crops. *Leucaena leucephala* or the 'ipil-ipil' tree was used by the Aztecs, Maya and Inca for inter-cropping with maize, and it travelled to Asia on Spanish galleons. In recent years it has become a favoured tree for intercropping because of its high nitrogen-fixing properties and its rapid growth. It is also used for applications in hilly areas, such as the tea gardens of Sri Lanka where, in addition to its other benefits, it retards runoff and provides shade to the tea bushes.

Starting in the 1980s, a CARE agroforestry project in western Kenya began distributing several million *Leucaena leucocephala* seedlings. Working closely with farmers and with the Kenya-based International Council for Research in Agricultural Forestry, CARE used schools as a starting point for introducing the concept to a community. Like most other primary schools in Kenya, Nyabeda Primary School in Siaya District received government support in the form of salaries for teachers. Everything else, including buildings, supplies and upkeep, had to be provided by the community, by the parents of the school's 600 children, or by the school itself. Some years earlier, the school had started a one-acre maize plot for provision of a daily hot lunch during the school year. Sales of the surplus were used to fund small school projects, such as roof repairs or a new classroom.

CARE began to provide support, training one of the teachers, Dishon Omondi, in inter-cropping and seedling production. A seedling plot was established near the school, and on the school farm, maize was intercropped with *Leucaena*. Five years after it began, Omondi ascribed several benefits to the project: the *Leucaena* provided fuelwood and building poles which previously had to be brought by children from reluctant parents; the surplus produced income for the school. Maize yields had improved slightly, but children no longer had to steal manure from home for the school plot, as the green manure from the *Leucaena* and the tree's nitrogen-fixing properties had taken its place. The community was happier, and a by-product was the school's ability to provide some parents with seeds and seedlings. Except for plastic bags for seedlings, Omondi saw little further need for CARE, and could point across the compound to a new 1.5 acre plot that he was starting, using the same techniques. It could be said that the project had become *sustainable*.

CARE saw the success in a different light. Although the technology had become established, the school had not yet worked out the planting and harvesting seasons correctly, because weather patterns were slightly different from those in other parts of the division. Little work had been done by CARE in documenting the quality and quantity of output – whether in green manure or maize. Clear economic analysis and documentation would be necessary if more individuals, NGOs and government workers were to accept the important extension issues that CARE was addressing. Nyabeda Primary School would require more attention in the next few years if the achievements were to be consolidated and properly replicated elsewhere.[16]

In conjunction with the Pan-American Development Foundation, CARE developed a similar project in Haiti, which in its first four years exceeded its target of three million seedlings by a factor of four. In Haiti the reason for success was not so much an acceptance of the full concept of agroforestry, as the fact that the trees had genuine economic value, serving first as a form of savings, then as a means of cash derived from timber sales when the need arose.[17]

A direct relationship between planting trees and carbon emissions from the burning of fossil fuels was made in 1989, when Applied Energy Services (AES), a Connecticut power producer, sought a means of offsetting 15 million metric tons of carbon dioxide to be released from a new coal-fired power plant. As the result of a World Resources Institute recommendation, AES funded a CARE agroforestry project in Guatemala, not only because of its offsetting environmental value, but because of its contribution to community woodlots, agroforestry, alley cropping and soil conservation. The project was also expected to benefit an estimated 40 000 Guatemalan farm families.[18] The Netherlands took a similar approach to offsetting the negative environmental effects of a new Dutch coal-fired power plant, by establishing projects in five Latin American countries to plant 125 000 hectares with trees.[19] Given the cost and the potential value of these initiatives, it is perhaps ungracious and unfairly rhetorical to ask why the South and Southern lands (rather than the North and Northern lands) should be used to offset increased Northern pollution. In the end, of course, the solution lies not so much in offsetting measures, as in the absolute reduction of carbon dioxide emissions.

'Minimum tillage' or 'zero tillage' farming is another farming innovation, developed at the International Institute for Tropical Agriculture (IITA) in Nigeria. In much of West Africa, only a fraction of the arable land is suitable for mechanized farming, leaving the rest to traditional hoe and cutlass techniques. Once cleared, however, land which for centuries has supported giant stands of hardwood and dense rainforests is sometimes barely able to support a crop of maize. In order to avoid disturbing already weak soil, 'minimum tillage' begins with a herbicide to kill the weeds. Seeds are then planted directly through the weed and crop residue, which serves both as mulch and as a further weed retardant. Despite the need for fertilizer and herbicides, there are major offsetting advantages. In an area where traditional family farming is back-breaking work, the minimum-tillage approach allows for increased food production without adding labour, and without the ravages that have been associated with expensive mechanization projects of the past. By reducing weeding, labour requirements drop dramatically, and erosion is minimized. IITA studies have shown that planting maize on a 10 per cent slope under minimum tillage can reduce erosion to one-fortieth of that following normal practice, and even on a 15 per cent slope there is virtually no erosion.[20]

Although logging for timber does not have to lead to deforestation, selective logging in the tropics is relatively unknown. In a 1988 study of forest management in 17 countries, the International Tropical Timber Organization found that only 800 000 hectares of forest were being maintained sustainably. A further 3.6 million hectares were being partially managed in India, bringing the total to 4.4 million hectares.[21] This represented about a half of one per cent of the world's remaining productive tropical forest. Since then, however, there has been considerable movement. Agenda 21, the Earth Summit's plan of action, provided detailed guidelines for combating deforestation. The Forest Stewardship Council, formed by environmentalists, timber producers and certification bodies in 1993, has developed principles and criteria for sustainable forestry, and provides a 'good housekeeping' certificate for timber and traders that meet established criteria. While only three per cent of the wood traded internationally in 1996 was certified, it was a promising start. And in 1997, the World Bank, along with environmental and business groups, endorsed a plan to raise the area under certifiably sustainable forest management to 200 million hectares by 2005.

One of the problems in sustainable forest management is the lack of a suitable technology that would allow selective logging. In order to get the logs out of a forest area, heavy equipment is almost always required. Roads must be built, and as a result it is uneconomical not to take as many trees as lax government supervision will permit. Even where it is practised, selective logging damages the residual vegetation because of the heavy equipment used to cut and move logs. An experiment in selective logging in Surinam found that 9 per cent of the remaining trees were destroyed and a further 16 per cent were damaged.[22] A similar experiment in the eastern Amazon found that while only 3 per cent of the trees were removed from a particular area, another 54 per cent were uprooted or damaged during roadbuilding and in the logging operation itself.[23]

Ultimately, demand reduction, policy intervention and enforcement are the answers to these problems, but technology can help. In Papua New Guinea, where the forest is seen more as a source of income than as potential cropland, a simple piece of appropriate technology has made a difference. The 'Wokabout Somil', a sawmill produced originally by a locally owned enterprise in Lae, requires no roads and no vehicles.[24]

Essentially, the Wokabout Somil consists of two blades driven by an 18 horsepower petrol engine. The engine is mounted on a chassis-and-rail assembly that straddles a log lying on the ground. It is the blades that move, rather than the log. The Somil is easy to dismantle and move about the forest, and the logs – cut into timber – can also be transported manually. The Wokabout Somil, of which more than 450 are now in operation, opens new opportunities for local processing industries – in building trades, carpentry and furniture manufacture. And in the hands of people who control a forested area, it offers a unique tool for forest management. In Bau, for example, a village in Morobe Province, a Somil was purchased in order to create employment for school leavers. The village realized that conservation was as important as income, and of the 6000 hectares it controlled, 3000 hectares were set aside for conservation. Knowing better, then, what the limitations and potential were, the villagers realized that the forest could support another Somil, and two were soon operating, each on a multi-shift basis.

Animal husbandry

As with agronomic research, work on poultry and livestock has reflected the priorities of the larger farmer, and of the urban demand for eggs, milk and beef. Exotic cattle have been introduced at the expense of research on local varieties, and much work on animals that are the priority of the poor – goats, sheep, pigs and chicken – has focused on improvements in 'factory' settings, rather than improvements that might benefit the very small farmer. Yet the productivity of animals has a greater importance for the poor, whose livelihood or even survival may depend entirely on the health of a single ox or a few chickens.

Low-cost techniques can be applied to animal husbandry and poultry. In Bangladesh, BRAC began a poultry-rearing project in 1985 with destitute village women – 'destitute' being defined in most cases as abandoned, widowed or divorced, landless, and, in most cases, head of household, with two or three dependents. Starting on a pilot basis, BRAC trained different women in three sets of skills: the first groups were trained in vaccination and the treatment of simple poultry ailments. The second groups, with one cock and nine hens each, were designated as trainers. Members of a third type of group, 'model rearers', were supplied with 3 cocks and 30 hens each. They were to supply others with eggs for hatching. The cost of the chickens and vaccine were covered by loans taken by the women from BRAC's savings and credit programme, and recovery rates exceeded 95 per cent. From small, experimental beginnings in 1985, the poultry programme had trained 41 228 women in poultry vaccination by 1997, and 1.2 million women – supplied with almost 10 million day-old chicks that year – were involved in poultry production.[25]

A similar approach to livestock development was undertaken by ITDG in Kenya.[26] Livestock is an important economic activity for large numbers of people, and is virtually the only source of income for the 25 per cent who live in arid areas. Although there are 11 million cattle, 6 million sheep and 8 million goats in the country, production of milk and meat is far outstripped by demand, with poor performance blamed on low producer prices, increasing costs of inputs, a shortage of credit, and livestock diseases.

Government veterinary services have been growing, but they are far outpaced by the need. In Nginyang Division for example, each of the 13 veterinary personnel are responsible, on average, for 324 households spread over an area of 342 square kilometres. Only one of the vets had access to transportation. In Tigania Division, each of the veterinary personnel served 2000 households in a 50 square kilometre area, and in Tharaka, an officer theoretically served 1200 households and an area of 187 square kilometres. Needless to say, those farmers closest to the veterinary station were serviced; the vast majority were not. As a result of such poor coverage, it is conservatively estimated that Kenya loses over $43 million annually through preventable livestock disease, with eight conditions accounting for 80 per cent of cattle disease and five complaints accounting for 90 per cent of reported disease among sheep and goats.

ITDG surveys indicated that as many as 33 per cent of the cases of cattle disease and 40 per cent of the sheep and goat ailments would not require a vet if farmers were trained in simple preventive and curative techniques. Working with Oxfam, the Kenya Freedom from Hunger Council and other local organizations, ITDG developed and tested training programmes and follow-up courses for farmers who were nominated by interested villages. The 'barefoot vets' received their incentive in the form of small fees charged to the users and through the establishment of a shop which stocked veterinary supplies and drugs. Preliminary results were encouraging. In one district, 27 men were trained as barefoot vets and 24 women were trained to vaccinate chickens against Newcastle disease, a major problem in the area. In a review conducted after the first 18 months of the programme, it was found that about 6000 chickens had been vaccinated and 4600 animals treated. Half the treatments were for one of the most simple of ailments – worms – yet it had been a costly one that people in the area had known little about.

The Turkana rainwater harvesting project

Brief highlights of successful projects can do two things. They can show how a different approach to development, and appropriate technologies, can make a difference. They can show that small, simple, cheap and non-violent approaches are possible, replicable and sustainable. They can also be misleading, for with each of the examples mentioned above, there were discouraging setbacks and failures before success was achieved. For that reason, it is worth examining a second ITDG project in Kenya in greater detail.[27]

The Turkana people, about a quarter of a million in all, live in an arid part of north-west Kenya that is about 50 per cent larger than Switzerland. They are a pastoral people, depending largely on the herding of cattle, sheep and goats, and on small sorghum gardens planted mainly by women during the rainy season. Raiding — attacking other groups in order to expand their cattle herds — has been a common feature of Turkana life for generations. Although officially 'subdued' by the colonial government in 1917, raiding continued, and in 1980, Turkana forays into Uganda returned with more than cattle. They returned with contagious caprine pleuro-pneumonia and rinderpest.

The diseases, combined with a particularly harsh dry season, were devastating. In a year that the Turkana called *Lopiar*, which means 'the sweeping', 90 per cent of their cattle died, along with 80 per cent of their small stock, 40 per cent of their camels and 45 per cent of their donkeys. Within a few months people were no longer able to feed themselves, and began to flock towards the few small towns in the area. Some starved; cholera broke out; some committed suicide. An EEC delegation visiting the area in 1980 estimated that there were 27 000 completely destitute people in feeding centres, and countless others in severely reduced circumstances. By 1982, the numbers had risen to over 50 000.

Famine was no stranger to Turkana. A 1961 drought had brought Oxfam and other NGOs to the area. They had promoted the resettlement of destitutes on the shore of Lake Turkana, and provided them with gill nets and canoes. Initial success in exporting dried fish to other parts of the country encouraged NORAD to assist in the building of a paved road into the region. And in order to develop the fishery further, NORAD also constructed a fish-freezing plant. Before it was finished, however, the fragile ecosystem of the lake began to react to the new industry, and catches plummeted, dropping from a high of 17 000 tons in 1976 to only 600 tons in 1985. The freezing plant was never commissioned and many families returned to cattle. Other pre-1980 projects, involving church organizations, FAO and UNDP, contributed to the development of various irrigation schemes covering approximately a thousand hectares on two Turkana rivers. A NORAD study found that these projects were unable to provide participants with enough income to cover even subsistence needs, denouncing them as 'disguised famine relief camps'.[28] Other studies found that at $61 000 per hectare, the Turkana irrigation projects were amongst the most expensive in Africa. (Even World Bank irrigation projects, rarely the least expensive, seldom rose above $10 000 per hectare in the 1980s, and usually averaged between $1000 and $5000.)

Nevertheless, when *Lopiar* struck, irrigation again came to the fore. The reasons were simple. First, although the previous schemes had been prohibitively expensive, it was thought that more appropriate technologies could do the same for significantly less. Secondly, pure relief without some sort of end in sight contradicts everything aid agencies believe in. And third, nobody (except the Turkana) was interested in cattle.

So huge amounts of food were trucked in, and people were employed in food-for-work schemes, building bunds and levelling gardens that could produce sorghum during the rainy seasons. Over 100 kilometres of contour bunds and embankments were built in the first three years, but construction was so poor that many failed to collect water, and others were simply washed away with the first rain. In order to limit the psychological damage caused by the disaster and to separate the hoped-for development aspects of the programme from what had become a thinly disguised relief programme, a separate 'water harvesting and draught animal demonstration project' was established, funded by Oxfam and supported with technical assistance from ITDG. Adrian Cullis, who had been a Voluntary Service Overseas worker, who had worked with the Salvation Army in the same area between 1979 and 1981, and who had spent time in the Negev Desert studying Israeli techniques of rainwater harvesting, returned in 1984 under ITDG auspices. He introduced several innovations the following year. The failed contour bund was abandoned in favour of a trapezoidal bund, and workers were taught basic levelling techniques with spirit levels. A demonstration herd of donkeys was established. Using a Swiss three-pad collar made from local materials, they were put to work with locally designed scrapers, earth scoops and an Ethiopian-style plough. These were popular innovations because the levelling of a garden and the construction of a single bund could take 15 people as much as seven months of hard work.

During 1985, four gardens were completed in time for the rains, but only two produced a harvest. The results were not as good as they might have been, but they did not create the negative impression of earlier efforts. People were prepared to continue working. Then, when Cullis was away from the area recovering from an illness, five of the donkeys died. Project staff had been unable to prevent people from overworking them once the foreigner, whose property they were deemed to be, had left the site. Then reductions in the food-payment allocations resulted in many of the men going back to cattle, leaving women to work for whatever they could get in the gardens.

Nevertheless, before the 1986 rains, a total of 18 gardens were prepared, and good crops were established. That year it was decided to discontinue food-for-work altogether as a means of getting bunds repaired. People would have to do it themselves, or they would fail. Left to make their own decisions, families at last decided to use their own donkeys for draught work. By 1987, 50 gardens were prepared and the rains were good. But once again disaster struck. Because an external variety of sorghum seed had been available at half the price of the better Turkana variety, most gardeners had chosen the cheaper seed, and yields were poor.

Encouraged by growing local interest in the project, even though the gardens were not producing what had been anticipated, Cullis and Oxfam decided that greater responsibility should be devolved to the garden groups. These groups, in fact, were based on a strong community spirit among the Turkana, and on a traditional system of governance which required careful discussion with the community's elders of any change or new ideas. In 1987 local organizing committees were set up in the three gardening areas of the project, and in 1988 a Management Board with committee representatives was established to run the project. Cullis remained involved as an advisor, but key decisions were now made by the committees and the Board. In 1988, the first year of self-management, 104 gardens – about 20 of them owned and operated exclusively by women – were planted. And 95 had successful harvests. Although some of the harvests were small, the best recorded a remarkable 1300 kilograms per hectare. In all, over ten tons of sorghum were produced, along with smaller amounts of cowpeas, maize, greengram and watermelon.

On sustainability

In 1997, ITDG sent an evaluation team back to Turkana to look at the impact and sustainability of the project. They found 342 working gardens. Thirty others had been abandoned – mainly by people drawn to the project by the initial food-for-work relief. Yields in years of good rainfall were similar to those elsewhere in the area, but in years of poor rainfall, they were significantly higher. The main beneficiaries of the project were women, whose control over the use of the harvest and whose status as providers for their families had improved. Better sorghum harvests had made a positive contribution to their families' diet and to their income, as evidenced by increased livestock holdings.[29]

Overall, the Oxfam/ITDG project demonstrated a number of important lessons. It had shown that it was not necessary to spend the $61 000 per hectare that had gone into the projects of the 1970s; the original Oxfam/ITDG schemes cost between $625 and $1000 per hectare.[30] The project also demonstrated that it was important to get the technology right. Small contour bunds, no matter how cheap, were too expensive if they did not hold water. It showed that success is not a simple process of discovery and action; it results from planning, action, reflection on the initial results, and further planning. It is also the result of a long process of dialogue with project beneficiaries and community leaders, and of building on their own social organizations. The improvements worked best with those already involved in sorghum production, building on existing skills and knowledge. It did not work well with people who came to the project from other backgrounds. Success came from encouraging success, but it also from recognizing and learning from mistakes; learning from the reactions of people to challenges and failure. This underlines two of the most ignored and yet most badly needed things in development work: *the need for a premium on documenting and understanding failure*. If mistakes are not understood, and if they are not remembered, there is every likelihood that they will be repeated.

The project demonstrated that when forced to begin taking financial responsibility for something with potential, people will do so. It also demonstrated that when they are made responsible for their own actions, people can do the right thing. The committees and management board did not abuse the trust placed in them by the donors.

The project demonstrated the importance of time and continuity to success. It had taken eight years for the first significant positive achievements after *Lopiar*. Some of the success was due to the pre-1980 efforts, from which the project was able to learn. But the persistence of Oxfam and ITDG in supporting the project, despite continued apparent failure, was critical. Part of the success was also due to something that few aid projects enjoy, the continued presence of a project *champion*. Adrian Cullis's experience of rainwater harvesting in Turkana spanned more than a decade, as volunteer, as an ITDG field worker, and finally, in the early 1990s as a visiting adviser and friend.

The reality of a dozen hard years in Turkana puts a new perspective on the definition of agricultural sustainability. Conway and Barbier define it as 'the ability to maintain productivity, whether of a field or farm or nation, in the face of stress or shock.'[31] The stress might be small or large, temporary or permanent. It could be the result of local factors, such as drought, flood or grasshoppers. Or it could come from external forces such as an increase in the price of fertilizer or the withdrawal of technical support. 'Sustainability thus determines the persistence or durability of a system's productivity under known or possible circumstances.'

Time is a key element of success. Many projects are based on unrealistic expectations and are too dependent on the standard development agency time-frame of two or three years. Some efforts can take hold very quickly, as in the case of the Wokabout Somil. Others take longer. Good management and 'self-help' are obvious examples of this. People will respond if they see value in something. The opposite has proven to be the case in almost every social forestry project where the external priority – getting trees into the ground – took precedence. Such projects usually fail. But where trees have been tied to income generation or real and visible community benefits, as at Nyabeda Primary School, projects have succeeded.

And any project having to do with land is far more likely to succeed if people own the land they farm or have secure tenure, than if they do not. The structure of land ownership is the most important determinant of the distribution of rural income and wealth, and in most countries it shapes the entire character of rural development. In Asia, the positive relationship between security of land tenure and increased productivity has been amply demonstrated, as has the viability of small, owner-cultivated farms. Farmers not only work harder on land to which they have secure rights, they are considerably more willing to take risks and to make the higher investments demanded by high-yielding varieties of rice and wheat. Losing one's land, on the other hand, has fundamental and devastating economic and social implications, including increased vulnerability to debt and coercion, and loss of self-esteem.

Much has been learned about rural development over the past century. Lessons, however, are only useful if the level of integrity in monitoring and evaluation is high, and if the lessons are remembered and applied. Many 'lessons learned' in the 1960s were ignored or forgotten in the following decades. The lessons illustrated in Turkana and in thousands of similar projects around the world relate to project design, target groups, cost and technology; to external factors, management and political considerations. Briefly stated, they include the following:

- rural development projects are most successful when they are low-cost and small in scale, when they respond to the needs of a specific target group and involve the beneficiaries themselves in the planning and implementation process;
- there must be a thorough knowledge of the project area – its people, the power structures, the role of women; readily available, conventional data are rarely adequate;
- projects must have adequate lifespans; planning and scheduling should be flexible, and must be appropriate to the often unstable environment in which they are located; projects must have a capacity for self-correction;
- rural development must be built around well-defined, appropriate technologies and methodologies *that become available in an appropriate sequence*;
- projects must anticipate relatively low levels of formal education in rural areas, and should be cautious in their use of high-cost technical assistance. Projects are more likely to succeed if they build on existing local capabilities, than if they attempt to impart completely new skills and knowledge.

Part of the original definition of 'sustainability', therefore, remains directly relevant. 'To sustain' still means to withstand, to maintain, to prolong, to be supported from below; to keep up courage; to affirm the justice or validity of the endeavour.

One of the greatest errors that donor agencies make in approaching rural development, is in defining sustainability in terms of their own assistance, and then in unrealistically constraining that assistance. Donor agencies regard their interventions as 'catalytic' – bright ideas which, combined with capital investment, will spur a group or an area onwards into the bright sun-lit uplands of sustainability. Then, in a specified period of time – often only a couple of years – having achieved sustainability in one place, the donor can move on to the next. Often the first question asked in a project approval process is, 'How soon can we withdraw from the project?' One of the worst innovations in project development during the 1990s, in fact, was the 'exit strategy', often required before the 'entry strategy' has been fully understood or articulated.

Life, of course, does not work this way. If development flowed so easily from good ideas and time-bound capital investment, the Indonesian and Brazilian rainforests would be intact and the Turkana would be living happily on the proceeds of frozen fish exports. Development, especially for poor people living on poor land, has a colder, harder reality to it. Sustainability may require longer-term external inputs; it may require less dogmatism about things like market prices for fertilizer in all places and in all countries. And it may require donor support for the recurrent expenditures that a government, with a weak economy and pressed by too many donor-inspired 'catalysts' in too many places, simply cannot meet.

Politicians and civil servants are usually judged on performance *now*, not in the future. The size of this year's grain harvest, therefore, is often judged to be more important than longer-run development issues. But planners have begun to see that it is as important to ask who is producing the grain, how it is being produced and where, as it is to know how much has been produced. The where, who and how will help answer questions about how much, when one looks towards the future. In determining what sort of assistance is required, and over what period of time, sustainability will then be viewed not only in terms of agricultural production, but in terms of sustainable incomes *based* on agriculture.

CHAPTER VIII
Post-harvest technologies

There are no successful projects, only those with less problems than others.
A.O. Hirschman

POST-HARVEST TECHNOLOGIES and food processing are important aspects of food security, and they are essential links in the food chain. Improved food preservation and processing techniques can reduce losses and increase yields, thus increasing net farm output and family income. They can retain greater value-added in the family or the village, rather than elsewhere. Traditional oil extraction techniques, for example, are time consuming and are often very inefficient. Simple improvements can increase yields, reduce workloads – often for women – and increase earning potential. Scaling up from traditional techniques offers one type of advancement, but so does scaling down from more modern technologies. This chapter examines one of the oldest food preservation techniques known, drying, and shows how the economics of scale can be the most critical factor in change. It also shows how scaling down – in Kenyan sugar production and Zimbabwean sunflower oil production – can save money and create jobs in the right sort of economic climate. Examples from Ghana and Tanzania demonstrate opportunities and problems in scaling up vegetable oil production.

The cost of drying out

Drying is the oldest method of food preservation, and it remains the most common.[1] For centuries it has occupied the imagination and inventiveness of entrepreneurs, innovators and food technologists. Gail Borden, whose name today is almost synonymous in North America with milk, had a passion for drying food. He spent six years marketing an unpleasant 'meat biscuit' before inventing milk powder during the 1850s. The meat biscuit is forgotten, but Borden is credited with helping to win the American Civil War for the Union by keeping its armies fed.

Drying retards the growth of bacteria and fungi during storage, and in the case of grain, prevents germination. With the development of new crops and new varieties of wheat and rice in the 1960s, new methods of drying have become a means of reducing loss and, therefore, of improving incomes. For example the introduction of irrigated rice crops during the dry season in India and Bangladesh meant that for the first time, rice could be harvested during the monsoon. Traditional drying techniques, therefore – spreading rice on the ground in the sun – were no longer as effective as they had been.

Much effort has gone into the development of low-cost mechanical and solar driers, but through a process of trial and error, ITDG learned that there are limits to the potential for scaling up from a traditional technology. Much of the drive behind the introduction of improved drying technologies and better storage facilities has to do with post-harvest losses, which in some of the poorest countries are routinely estimated at 10 per cent, and are sometimes as high as 40 per cent. If such losses could be reduced, it stands to reason that both farmers and the national economy would benefit significantly. Economist Martin Greeley has disputed the received wisdom on post-harvest losses, however. He argues that they are often exaggerated, and that the 'uncritical acceptance of convenient but unfounded guesses' by 'experts' has helped justify huge technology-driven aid programmes aimed at selling large grain-storage facilities.[2]

In Bangladesh, the distinction between two types of drying became apparent to ITDG in its development of mechanical driers. The first type of drying is for product preservation. The second has to do with product refinement. Small-scale driers, solar, mechanical or otherwise, have had little success among the poor when preservation is the issue, for several reasons. The first, as Greeley argues, is that losses have been exaggerated. This is borne out not only by research findings, but also by the observable and historical fact that farmers and consumers vote with their money. The costs associated with new loss-prevention technologies, even small ones, are in many cases simply not justified by the benefits. Because the bulk of the food that poor people produce is for home consumption, there is no cash benefit derived from preservation, thus providing no financial motivation, either family-based or communally, for a new drying or storage technology.

The situation is different, however, where there is a marketed surplus, and where the price of a product is sensitive to quality increases. Dried apricots, sold throughout Pakistan, represent a significant source of income for villagers in the remote northern areas where apricots grow. Traditional sun-drying techniques are adequate, but the introduction of a simple solar drier produced locally of wood and polyethylene sheet, and a sulphur smoking process, make the product more hygienic and improve its colour. It is the improved colour which has added significantly to the product's appeal and to its price, justifying the small additional cost to producers of constructing a drying tent in the village.

It is in this 'product refinement', the processing of a raw product from one form into another, that newer drying technologies have the most potential. Unfortunately, as with so much modern technology, the gap between what is technically and economically available to villagers in remote areas is highly limiting. Modern technology has produced large-scale spray and freeze-drying systems that cost tens of thousands of dollars, and more. But what Fritz Schumacher called 'the missing middle' – an *intermediate* technology – is as painfully obvious in this technology as in others.

The 'product refinement' message has become most evident through experiments conducted by ITDG and others in Latin America and the Caribbean. Initial experiments to develop a drier in St Vincent, for the production of dried sorrel – used in a traditional drink – were conducted under a British aid programme. The individual who carried out those experiments further refined the technology on a subsequent assignment in Guatemala. There, a Swiss development organization, working with poor villagers in the highlands, investigated the possibility of agricultural diversification. The villagers' land holdings were too small to produce enough beans and maize for subsistence, and it was thought that higher value vegetables, produced for sale, might ease their economic plight. The Instituto de Nutricion de Centro America y Panama (INCAP) knew that packet soup manufacturers were importing dried vegetables from Europe for the Central American market. Further research showed that 14 000 kg of parsley alone was being imported, at a price of $8 per kg, almost twice the price of the next best vegetable crop known to the villagers.

The drier that was eventually produced was a simple plywood cabinet containing two chambers with nine trays each, served by a thermostatically controlled heater and blower. Washing and sanitizing systems were developed in order to meet microbiological requirements, and a semi-continuous mechanism also evolved. Because the parsley at the bottom of the chamber dried first, the trays had to be constantly shifted. A simple lever mechanism solved the problem, sliding the bottom tray out and making room for a new tray at the top.

One of the INCAP students, returning home to Peru, asked ITDG for assistance in developing a prototype there. A Peruvian herbal teabag firm, which had been importing its raw material, found its sources cut off by customs restrictions, and had begun to experiment with dried camomile produced in the Andean region. ITDG had made an agreement with the Peruvian firm stipulating that the technology, if successful, would be freely available to other interested parties, and the first to come calling was Julian

Dolorier, a small equipment manufacturer whose firm in Lima had already produced three tray driers on its own for dye fixing in cloth. Using the ITDG technology, Dolorier sold 69 tray driers between 1985 and 1996, mainly to small firms in Lima and other urban centres. Over time, a series of improvements helped him reduce the cost of his driers, enabling the company to remain competitive in the market-place when import restrictions were removed in the mid-1990s. Dolorier's driers were improved, installed and maintained without any external assistance beyond the development of the early prototypes and an information package produced by ITDG. In addition to camomile, verbena and lemon balm, the new driers were used for fruit, vegetables, rice, coffee and the fixing of dyed cotton cloth.[3]

The success of the technology, which has also been taken up in Colombia and Cuba and which shows promise elsewhere in Latin America, was based on several factors. It was cheap, simple and locally adaptable. It produced dried fruit and vegetables more economically than modern imported tray driers. Equally important, however, the drying of the herbs and vegetables in the successful Latin American cases was only part of a process which started with the growing and ended with packaging (teabags, herbs) or further processing, as in the case of the Guatemalan soup mixes. In other words, the driers became part of a process that already existed; they were a technological and economic refinement.

Scaling down: sugar

There are six basic steps in the production of sugar. First, cane is crushed and juice is extracted. The juice is then clarified, boiled, crystallized and 'centrifuged' to separate the solid sucrose from molasses. Then it is dried and packed. By the end of the eighteenth century, a more or less standard technique had developed, boiling the juice in large open pans and passing the fumes from burning sulphur through it, while adding enough lime to counteract acidity. This 'open pan sulphitation' (OPS) process, however, had drawbacks. It produced brown sugar of uneven quality, and the fuel required for the boiling process was expensive. Normally, heat was derived from burning the cane residue, or 'bagasse', but this usually had to be supplemented with wood or coal.

By the middle of the nineteenth century, a new process had been developed and commercialized. It revolved around the creation of a partial vacuum, which caused the sugar to boil at much lower temperatures. This had two effects. It produced a more uniform, lighter sugar. And the fuel required for the heating process could be derived entirely from bagasse. Other improvements, such as higher extraction rates in the rolling process, improvements in refining, and the automation that came with steam power, combined to make this 'vacuum pan' (VP) process the favoured technology.

But there are several reasons why the older OPS system has clung to life, and these explain ITDG's unlikely involvement with sugar, first in India and then in Kenya. The much more expensive and more sophisticated VP system requires high levels of managerial talent, and in most countries the equipment must be imported. The most efficient VP systems consume between 10 000 and 20 000 tons of cane per day, far exceeding demand in many countries. For countries with limited demand, smaller operations would make more sense, especially if the trade-offs between local production and continued import are clear. A 1986 review found 26 countries, 21 of them in Africa, which consumed less than 150 000 tons of sugar annually, and which were net importers.[4] For them OPS has genuine potential. Other countries have a demand that might justify a large-scale plant, but they do not produce adequate supplies of sugar, and are thus obliged to import. But one or more small factories could use even limited amounts of local cane, thus reducing imports. In addition, for poor people in marginal farming areas, small local plants could encourage the introduction of a cash crop and the jobs associated with it.

It was these possibilities, and the continued existence of small OPS factories in many countries, that enticed ITDG into the field in the late 1970s. In China, there were an estimated 169 sugar factories with capacities of less than 500 tons per day. In Japan there were 13 with capacities ranging between 300 and 1000 tons, in Costa Rica there were 12, and in India, 35. Much smaller versions in India numbered in the thousands. The OPS process had actually *re*-emerged there in the 1950s in response to an oversupply of cane and the attractive economies of this system.

There were major problems of efficiency, however, which – in a climate of control and fixed prices – threatened the viability of the technology. The Appropriate Technology Development Association (ATDA) in Lucknow became involved, and with ITDG assistance made two essential improvements to the technology. The first was the development of a screw cane expeller which raised milling efficiency from 75 per cent to 85 per cent. The second was the introduction of larger boiling pans and fuel-efficient furnaces. This had the effect of more efficient and faster boiling, which reduced fuel costs and sugar losses.

ITDG's interest in the technology was predicated on its potential for sugar-deficit countries with a tradition of smallholder cane agriculture, one of which was Kenya. There, in the mid-1970s, two OPS factories had opened and closed in short order. ITDG made an arrangement with a third, the West Kenya Sugar Company (WKS), a private firm located in Kakamega, which had opened in 1981. Using Indian technology, the WKS factory had a capacity of 200 tons of cane per day. Its owners had been encouraged by a favourable investment climate, and by the 800 small farmers near the factory. But the problems that had shut the first two factories became evident very quickly. Because of the different qualities of local cane, crushing efficiency was lower than in India, and because the climate was significantly wetter, the bagasse did not burn as well, requiring greater supplements of fuel. In addition, the factory was located at an altitude of 5000 feet, which further reduced fuel efficiency. What had seemed economically promising now looked like a potential disaster.

ITDG's technical involvement revolved around improvements similar to those made in India. A better shell furnace, based on the ATDA design, was developed, which improved fuel efficiency and the utilization of capacity. An Indian screw expeller had been imported and tested by ITDG, but proved unreliable under Kenyan circumstances, and WKS made other improvements to the crushing process. Despite the innovations, however, by the late 1980s ITDG was beginning to realize that the problems of the sugar industry were not only technical in nature. ITDG had involved several economists in the project, and through comparisons with the larger VP factories in the country, it was becoming clear that further dissemination would be a business problem, rather than a technical matter.

Depending on prevailing prices, OPS sugar could definitely be competitive with VP sugar. Moreover, the capital investment in OPS per ton of crushing capacity was 60 per cent lower than that of the VP process.[5] Installed in marginal areas, it did add to farm income and create new jobs – 630 during the high season at the West Kenya Sugar Factory. But ITDG's interest was not so much in the success of the West Kenya Sugar Factory, as in the potential for wider diffusion of the technology. This depended entirely on commercial investors taking up the opportunity provided by WKS, and the opportunity remained unpromising. Because the large VP sugar factories were government-owned, they operated in an economic twilight zone. When the price paid to farmers for cane was increased by government fiat, they paid it. When the price of wholesale sugar was held down, they endured it. Original investments were written off. Taxes were postponed, avoided or ignored. Losses were absorbed by the owner (i.e. government), a business technique unavailable to the private investor.

At the time, the policy of ITDG and other AT institutions was shifting towards greater focus on facilitating the ownership of new technologies by the intended beneficiaries – the poor – rather than by surrogates who might offer them only employment. This shift may have been triggered, in part, by the growing realization that the sugar sector policy in Kenya, despite passionate government assertions to the contrary, was firmly biased towards the large-scale, capital-intensive investor. Believing in addition that further investment in sugar would move ITDG too far from its basic objectives, plans for further OPS investments in Kenya and Bangladesh were quietly shelved, and ITDG's interest was downgraded to an information and consultancy function.[6]

The experience demonstrates a fundamental problem in the development of appropriate technology. 'Small' is a relative term, but in the AT world it is often defined in absolute terms because of funding constraints. In other words, a small cow is still a cow. But with inadequate funds, even if you need cow, you may have to settle for goat. The OPS process is small in comparison with the modern VP technique, but it is not cheap. Even a small factory can cost $2 million. ITDG's investment in OPS, at roughly $500 000 over a decade, was enormous by NGO standards, but it was tiny in comparison with the scale of the problem, and with the scale of the potential opportunities. In the absence of an institutional investor – either commercial or aid-related – it was two NGOs, ATDA and ITDG, that shouldered the burden of research and diffusion. If they can be criticized for getting in over their heads, or for dropping out at the half-way point, they can also be congratulated for taking it as far as they did. Appropriate technology is not just limited to small investments because of philosophy, it is limited to small investments because those with money – governments, aid agencies, business and much of the research community – are very often wrapped up in their own vested interests.

Oil in troubled economies

Oils and fats are important nutrients, and they make starchy diets more palatable. For people in the North, a reduction of intake is usually warranted; the problem for the poorest people in the South – where cost and availability are serious problems – is how to increase it. Methods of obtaining oil from oilseeds date back to ancient Chinese stone mills and wedge presses, and to screw presses invented by the Romans for olive oil. Today in the South, vegetable cooking oil is derived from coconuts, oil palm, ground nuts, sunflower and shea nuts. Other seeds produce edible and non-edible oils for soapmaking, paints, lubricants and cosmetics, and include neem, rape and mustard, linseed, castor, cotton and sesame seeds. In many countries a traditional, village-based 'women's activity', vegetable oil production has in recent years been increasingly taken over by large-scale business interests.

Traditional oil extraction can be labourious and inefficient. Groundnut oil, for example, which is an important source of revenue for women in many parts of Africa, takes between four and five hours to produce. The groundnuts must be shelled, grilled and winnowed. Then, with mortar and pestle, they are ground into a paste. Boiling water is added until the oil separates; then the oil, skimmed from the surface, is in turn boiled to remove any remaining water. Palm oil production is even more time-consuming. After cracking the nuts by hand, women boil the fruit, let it cool overnight, and then pound it in a vat to which boiling water is added. The resulting liquid is boiled again to remove the water. The process, which takes several hours over two days, is so inefficient that there is usually a 50 to 60 per cent loss of oil. In addition to the hours spent directly on the oil production, women must spend time gathering fuel for what is essentially an energy-intensive process.

Ghana's Technology Consultancy Centre (TCC), seeking a means of reducing the price of palm oil for a soap-making process it had developed, began to experiment with a palm oil press in the mid-1970s. Made from a section of used pipe about 30 cm in diameter, a piston, moving on a large screw, gradually squeezes oil from the cooked fruit. In Ghana, this press had a significant price and performance advantage over models developed elsewhere. For example, oil presses produced in Cameroon and at the Royal Tropical Institute in the Netherlands used hydraulic lorry jacks rather than a screw. Widely available, jacks eliminate the problem of having the screw manufactured. In the 1970s in Ghana, however, there were already small machine shops capable of producing the screw from scrap truck axles, while a lorry jack was worth its weight in gold. Lorry jacks, intended for occasional lifting, also proved undependable when used every day.

One of the most important factors in the dissemination of the press was the involvement of the National Council on Women and Development, which made some of the original requests, tested early prototypes and placed the final versions with women in working village settings. The TCC press reduced the cost of palm oil production, but significant costs were still associated with the boiling process, and the press had not significantly reduced the overall time required. Gradually, therefore, one improvement followed another. Fuel-efficient boiling tanks, already being used in TCC's soapmaking process, were adapted for palm oil. And because TCC was working closely with small engineering workshops, an inexpensive palm-kernel cracking machine could be developed relatively easily.

The Ghanaian approach reflected the need for a technology that reduced time and cost. It succeeded because TCC listened to the users, and because it became intimately involved in the entire process of oil production, developing its approach in incremental steps as each new bottleneck or requirement revealed itself. And it succeeded in part because of the availability of light engineering workshops capable of cutting, drilling and grinding steel for prototypes and production models.

In Zimbabwe the situation has been very different. There, oil production has been centralized into the hands of four major producers in the commercial sector. Farmers who grow sunflowers suffer the vagaries of price fluctuation and see their product trucked away whole, to be returned as expensive, factory-produced oil in bottles with attractive labels. The challenge in Zimbabwe was not the same as in Ghana, where TCC advanced a traditional village technology to an intermediate level in order to save time and money, and to preserve jobs. In Zimbabwe the challenge was to move back a stage, to a less complex, affordable level of technology that could create new income opportunities in rural areas.

ITDG's efforts to develop a decentralized production capacity in Zimbabwe actually began six thousand kilometres away in the Indian city of Rajkot. There, a man named Veljibhai Desai had established a company called 'Tinytech Plants Private Limited' which produced complete packages of mechanized oil-producing equipment: decorticators to remove husks and shells, boilers, kettles, filter pumps, filter presses and expellers. Desai was a good salesman and in addition to the usual promotional material, he would send potential customers a two-hour video showing his machinery at work. He was also a man with a mission, a *champion* with convictions. 'All the dreadful problems appearing nowadays in the world, such as unemployment, hunger, exploitation, disparity, pollution, urbanization, war, etc., are direct result of heavy and centralized industries,' his brochure states. He foresaw a twenty-first century in which decentralized, small production units (some of which he would produce) would be the order of the day. 'Whenever I see any big factory or industry,' Desai continued, 'I am horrified to look at it, because it looks me as DEMON, threatening the extinction of humanity, its culture, its liberty, its freedom, etc. Small is possible,' he wrote, 'Small is inevitable.'

In addition to colourful brochures and videos, Desai produced good machinery. In 1987, following a two-week visit to Desai's factory and some of his satisfied customers in India, an ITDG food technologist decided that the Tinytech equipment might make sense in Zimbabwe. Desai had already sold a dozen of his plants to private companies in Tanzania and Malawi, and at about £3000 for a full set (f.o.b. Bombay), it was between one third and one tenth the cost of similar European or Japanese equipment. By 1990, a Tinytech oil plant was in operation, crushing sunflower seeds in Murombedzi, a small town two-hours drive north of Harare. Working with another international NGO which provided the local organizational base, ITDG focused on technical support and economic analysis of the operation.

On several levels, the preliminary results were excellent. In India, such a plant would probably run on a two-shift basis. But at Murombedzi, even on a one-shift basis, operating for eight hours a day, twenty-four days a month, a reasonable profit could be projected. Borrowing at commercial rates of interest and using existing prices for seeds, it was estimated that the mill could undersell commercially produced oil and break even at less than 50 per cent capacity output. Projections would obviously change with alterations in the farm-gate price of sunflower seeds and the retail price of oil. But essentially, the project bore out Desai's claim – that his 300 mills in India and Africa had proven small to be possible.

If the transfer of technology were that simple, however, Veljibhai Desai would certainly not need NGOs as his go-between. Investors would be clamouring to buy. In Zimbabwe, part of the reason they did not was that those with money to invest did not know about, or were not interested in, small-scale decentralized food production. And those who might have been interested had neither the money nor the know-how. Enter, stage left, the development-minded NGO – usually long on ideals and short on business acumen. The problems at Murombedzi started when ITDG's partner NGO began to demonstrate a casual attitude towards production procedures. As a result, the discipline of the eight-hour day that was necessary for economic production slipped.

But an even more elementary business error began to surface early in the production stage. The byproduct of sunflower oil – the residue – is an 'oil cake' which, when mixed with bran, can be used for cattle feed. Neither ITDG nor the partner NGO had made provision for selling or using the sunflower oil cake produced at Murombedzi. As a result, as oil production went ahead, oil cake began to pile up in the factory. It piled up and up. Besides having a limited shelf-life, oil cake can be a fire hazard, so when its physical presence began to impede production, Lever Brothers in Harare offered to buy the oil cake, partly because their more efficient machinery could extract a bit more oil from it, but also because they had a market for cattle feed. The partner NGO, having embarked on the enterprise in order to find an *alternative* to Lever Brothers' sunflower oil, rejected the offer. The oil cake continued to mount until, piled almost to the ceiling, work could no longer continue and the plant was forced to cease operation. Eventually an appropriate buyer was found, but there was a lesson in the incident: if a technology is to provide new employment opportunities for the poor, the choice of implementing partners and a clear understanding of *their* objectives and talents are as important as the choice of technology. If the partner – as in this case – is motivated more by ideology than commercial good sense, the project is unlikely to succeed.

At this point in the development of the Murombedzi oil mill, a chance editorial in a Harare newspaper bemoaned the shortage of cooking oil in Zimbabwe. Unwilling to let an opportunity for free advertising pass, ITDG wrote the paper a letter, describing the Tinytech operation. To its surprise, there was a flood of enquiries from commercial farmers, feed companies and aspiring investors. And so – unlike the sugar one – this project went into a second phase, aimed at testing its commercial viability more rigorously. Four small investors bought into the technology, and within a couple of years two had failed. But they failed for managerial rather than commercial reasons: absentee management, nepotism and attempts to compete with large firms in an urban market. In the other two cases, the results were little short of spectacular. The successful mills returned over 50 per cent on investment and made profits of 21 per cent on sales. A typical mill employed ten full-time and three part-time workers. People in the vicinity made extra money from the collection and sale to the mill of bottles and firewood; local machinists gained work from equipment repair. Farmers preferred to sell locally because they were paid immediately and in cash, while for consumers, the product was 20 per cent cheaper than oil from the major companies. An evaluation found that 1420 households (representing about 8000 people) gained in one way or another from each mill, and that the economic benefits per year were in the neighbourhood of Z$471 000 (US$51 000) – on a capital investment of less than Z$200 000 (US$21 600). By 1995, seven more mills had been established, and by the end of the century there were more than 60 in operation. Veljibhai Desai sold his mills throughout Africa, not least because – in the words of a 1995 evaluation – 'the Tinytech is a true intermediate technology in Schumacher's sense: it is non-violent, low-cost, a result of South–South transfer, and is an intermediate between traditional rural level technologies and conventional, industrial scale technologies.'[7]

The Bielenberg ram press

Similar issues and others emerge from the development of an oil press in Tanzania, named after its inventor, an engineer working for Appropriate Technology International, Carl Bielenberg.[8] The Bielenberg Press was not a variation or adaptation of known methods of oil extraction, it was a completely new innovation, developed in 1985 in the context of a village sunflower project jointly supported by Lutheran World Relief, ATI and the Government of Tanzania. Unlike the presses in Ghana, which use a screw to force oil from the seed, and unlike the Tinytech expeller which works on the shearing effect created by a wormshaft, the Bielenberg Press uses a small, hand-operated piston mechanism, or 'ram'.

Untreated sunflower seeds are fed into a hopper on the top of the press. In the vertical position, a long handle allows a measured amount of seed into a cylinder. The operator than pulls the handle back, activating the piston and forcing the seeds into a small cage. The holes in the cage allow the oil – now being squeezed out of the seeds under great pressure – to drop into a pan below. On the backward motion, the residue is ejected, and a new batch of seeds drops into the cylinder.

The press has several advantages in addition to its efficiency. It is robust, cheap (between US$105 and $280 in 1996, depending on the country and the manufacturer), it can be made locally, has few moving parts and thus requires little maintenance. It takes the entire sunflower seed without prior preparation or decortication. Because it is small, it can be moved easily, and unlike many screw presses, does not have to be anchored into a concrete foundation, because the weight of the person operating it holds it in place.

Within five years of its invention, the Bielenberg Press had achieved significant success in dissemination, with over 200 in operation in Tanzania, and experimental or production models in use in Zambia, Kenya, Lesotho, Mozambique and other countries of the region. In a sense, the success was almost too sudden – a case of a technology in advance of its time – for many of the ancillary requisites were not in place when it first appeared. The project was established in an area of semi-arid under-utilized land where there was both potential for increased sunflower production and a market for oil, but there were two production problems. First, sunflower was a new crop to many farmers. But even those who knew it had to be introduced to a new, lighter variety, because previously, sunflower had been sold to the government by weight, and so the heavier variety was preferred.

A second problem in dissemination revolved around ownership. In the early stages, Lutheran World Relief, through the Tanzanian Lutheran Synod, emphasized group, rather than individual ownership. The presses were made available to village groups and Lutheran congregations on a loan basis, with the idea that repayments would cover the upkeep of the project. Perhaps predictably, especially because no interest was charged, many of the loans were not repaid, and the financial viability of the project became questionable. Many of the group presses were used only sporadically, and efficiencies were low. But when presses were eventually made available to individuals, the situation changed dramatically; it was discovered that where communally owned presses produced, on average, 1000 litres of oil in a season, those operated by individual families or partnerships averaged 2440 litres.[9]

A third potential problem, manufacture of the machines, had been taken care of by the machine's simplicity, and by the development of simple jigs and fixtures which were supplied to manufacturers as part of a package to ensure uniformity. By 1989 there were three commercial manufacturers in the project area, and a commercial firm had started production in Kenya. With assistance from Africare and USAID, the Technology Development Advisory Unit at the University of Zambia modified the machine to suit local conditions, and began field trials with three dozen presses in 1989.

This sounds good, but from ATI's point of view, things were still not moving quickly enough. As with ITDG's Tinytech project in Zimbabwe, ATI decided to turn standard NGO ideas about ownership and dissemination on their head with the creation of a commercial firm – ZOPP Ltd. ZOPP's aim is to *sell* products and services throughout the region, based essentially on an improved Bielenberg, known as the Camartec ram press. By 1996 it was estimated that 5300 ram presses were in operation throughout Africa as a result of ATI's efforts, including 1500 in Tanzania.[10]

A fourth problem in the project's early days in Tanzania revolved around the question of oil price and availability. When the project was conceived, most oil was imported; prices were high and availability was a problem. By 1986, however, after the equipment had been developed and the cost of the technology was more or less fixed, import restrictions were lifted, and the price of oil fell. Fortunately this did not seriously affect the profitability of the press, but it signalled a problem frequently ignored in the development of appropriate technology. It is often price distortions and shortages in a troubled economy that make local production and new, appropriate technologies seem cost-effective. Liberalization or sudden changes in the price of inputs can change the situation drastically, often for the worse. When an analysis of the potential for a new technique or technology is being calculated, therefore, it is essential that provision be made for such possibilities where small-scale producers are concerned.

Although there were many reasons for the success of the Bielenberg Press, continuity in leadership, a factor often found in successful projects, was important. Not only were the Project Director and Field Manager present through the start-up and maturation periods, the project had access to the inventor of the press throughout its development. Dissemination of the technology beyond Tanzania had considerable support from ATI, through the production of technical manuals, reports, consultancies and workshops, and then through the creation of a commercial mechanism.

Doing it right

As with the development of the Ghana palm oil press, the Bielenberg press was like a pebble thrown into a pond. The ripple effect required agricultural extension work with farmers on new crops and seeds; the manufacturing technology had to be simplified and made available to commercial producers through the development of jigs and fixtures. The initial users of the press were inefficient, and a new approach to dissemination had to be developed. For the project managers, some of these ripples must have seemed like tidal waves that required exceptional swimming skills, but they were not dissimilar to the problems that have faced innovators from time immemorial. Those who can swim, or who learn how to swim, tend not to sink.

The Canadian International Development Research Centre (IDRC) frowns on this *ad hoc*, chaotic approach.[11] An IDRC specialist observed that the Tanzanian project was lucky to have had good seed when it was needed, and was lucky that farmers actually adopted it. The dissemination package, including agricultural research, manufacturing technology and financing, had all worked out reasonably well, but if one were to attempt to apply it on a national scale – in Kenya, for example – some 40 000 presses would be required to meet the annual oil requirement, rather than the 200 in the Tanzania project's first five years. It is perhaps wrong to dismiss good luck in the development of technology. One of Napoleon's most frequently asked questions about officers recommended for promotion was, 'Has he luck?'

But IDRC had a point. And the way it handled the matter illustrates the difference between the well-motivated, ill-funded NGO, and that of a well-heeled research institution, staffed by well-paid professionals with top academic credentials. Identifying a range of interacting components in the problem of increased oil production in Kenya, IDRC took a systems approach, beginning in 1987 with a consultancy that identified preliminary research requirements. A workshop on the findings drew together government officials and academics – both important to the long-term political and institutional viability of what might emerge.

Then, in 1988, Egerton University was engaged to carry out a comprehensive survey of the oil business, resulting in Phase One of a Vegetable Oil/Protein System Project to investigate the extent and magnitude of small-scale rural oilseed processing. Enthusiasm among Kenyan scientists, administrators and policy-makers led to a second, then a third phase in 1989, which had two aspects. The first aspect facilitated 'the integration and coordination of an evolving research programme', and the second consisted of satellite projects of applied research, such as the purchase and testing of a Bielenberg Press. This particular effort demonstrated, among other things, that the press was 'well received by the rural population visited, thereby creating an incentive for further investigation.'[12] This finding recalls Saul Alinsky's description of the University of Chicago's Sociology Department. It was, he remarked, a body which would spend $100 000 to discover the location of all the brothels in the city – information which any taxi driver could provide for free.

Along with its preliminary studies, findings and research data, IDRC produced a useful, comprehensive dissemination strategy for those interested in the Bielenberg press. The strategy could be modified for almost any food processing technology, and its elements are worth paraphrasing. In addition to detailed information on the technical and economic feasibility of the press, the strategy consists of two basic components. The first is an extension package for buyers or users, and starts with oilseed production technology.

Before investing, producers or extension agents need to know about oilseeds: where to get them, how they relate to other crops, and what they require in the way of land preparation, fertilizer, pesticide and weeding. They need to know about harvesting, and post-harvest treatment, as well as threshing and storage. Ram press users need to know how to operate and maintain the press, and about cleaning, packing and storing the oil and oil cake. Work may be required with consumers or potential consumers in the use and storage of oil and oil cake. Ownership of the press should not be clouded by unprofessional notions about groups. Groups may be able to take a technology and make it work, but the reason will not be because they are a group, it will be because they have the right mix of skills, discipline and ambition. Hard issues relating to marketing must also be addressed. These include questions of price, storage, packaging, labelling and promotion.

The second component of the IDRC approach related to the manufacturing, distribution and financing of the press itself – materials, workshops, quality control, prices and distribution. A financing system for both manufacturers and purchasers may be necessary for widespread dissemination, involving financial institutions and government organizations.

IDRC articulated the strategy that ATI and the Lutheran Synod had actually carried out, not in an orderly, theoretical fashion, but in a real-life situation. But like TCC in Ghana, they were small, and IDRC was thinking big. The contradiction remains, therefore, as it did in the case of Kenyan sugar: on the one hand, the 'amateurish' NGO, working against all odds to make a difference, sometimes successfully. On the other, a choice between the professional (and very costly) systems approach, or – as in the case of sugar – nothing. With only 200 in Tanzania, the Bielenberg Press did, in 1990, have a long way to go before making the impact that IDRC talked of. But by 1996, the numbers in Tanzania had risen by a factor of more than seven, and in Kenya – the IDRC study notwithstanding – there were only 250, most of them the result of ATI's work rather than IDRC's. IDRC had, in fact, left the field of post-harvest technology because of budget cuts, and all institutional memory of its oil press adventure had disappeared.

This may not matter much. In some cases the systems approach has had success, as with high-yielding varieties of wheat and rice in the 1960s. It is what accounts for the success of BRAC in Bangladesh, described in several places in this book. IDRC's persistence over twenty years in the development of a sorghum dehuller is another success story of this type. But in an attempt to replicate these successes, the few well-funded development research organizations may over-study things. They may be developing expensive templates for a real world that will never use or remember them, leaving poorly funded, ill-equipped NGOs to do the dirty work (perhaps commercializing themselves in the process), and then criticizing them for their limitations.

CHAPTER IX
Energy and power

Genius is one per cent inspiration and ninety-nine per cent perspiration.
　　　　　　　　　　　　　　　　Thomas Alva Edison[1]

THE OIL CRISIS OF 1973 fuelled an unprecedented interest in the development of alternative technologies. The 1979 oil 'shock' reinforced the drive. Within ten years, however, the price of oil had dropped in real terms almost to its pre-1973 level. Lulled into a somnolent sense of security that was accompanied by growing rather than depleting quantities of known oil reserves, Northern interest in energy conservation and in new, energy-reducing technologies flagged, came to life just before the 1992 UNCED Conference in Rio, and then flagged again.

In the South, the situation was somewhat different, in part because the shocks of 1973 and 1979 had been major contributors to mushrooming Third World debt. Countries now had to spend much more, simply in order to keep oil imports at a constant level. Between 1973 and 1980, for example, the oil import bill for non-OPEC African countries rose from $800 million to $8.4 billion. The lending capacity of Western banks, significantly enhanced by the vast increase in petro-dollars suddenly flooding Western financial institutions, fed the demand. African external debt rose from $10.6 billion in 1971 to $68.9 billion in 1982. Big money was badly needed, and big money was readily available. Combined with serious decreases in the value of primary exports, the long-term effect on most developing countries was devastating.

Confusing objectives

In the mid-1970s, when donor agencies began to take a greater interest in energy issues, they were responding in part to a desperate and collective Southern plea for assistance. It was a recognition that the oil crisis had destroyed the ability of many countries to maintain infrastructure, to sustain the development investments of previous decades, and in some cases, to service their debts. Combined with increased awareness of the global environmental effects of deforestation, support for alternative forms of environmentally friendly energy seemed a logical new direction for the development establishment.

Many of the projects, however, were far from successful, and a decade or so of patchy, often unprofessional activity left the South littered with windmills that didn't turn, solar water heaters that wouldn't heat, and biogas experiments that were full of hot air before they started. Developing countries and poor people were often used as guinea pigs for untested, unreliable gimmicks. It was a classic case of assuming that the 'transfer of technology' was simply a matter of invention, to be followed by widespread and unquestioning uptake among grateful consumers. Unlike other, more mature technologies, alternative energy seemed to attract more than its share of cranks. (A failed inventor is a crank; a successful one, however, is a genius. This recalls Schumacher's retort, quoted in his daughter's biography, *Alias Papa: A Life of Fritz Schumacher*, to the suggestion that he was a crank. 'A crank' he said, 'is a piece of simple technology that creates revolutions.') The ten-year love-in, which lasted from the mid 1970s to the mid 1980s, helped give appropriate technology and the concept of renewable energy an undeserved bad name.

Given how it evolved, sustainability of interest and funding was perhaps precarious from the outset, especially when so much was expected in such a short space of time. David Inger writes with frustration of the Botswana experience:

> The Botswana Renewable Energy Technology Project dropped out of the skies into Botswana in 1981 and disappeared four years later. In the interim, this USAID project did a lot of damage to the credibility of renewable energy and appropriate technology, and also to local institutions already working in the field. The project assumed that progress could be made in renewable energy technology development and dissemination in three to four years, and to make matters worse, it was badly managed.[2]

No less critical was USAID's 1986 evaluation of its own rural energy projects, which found that they had failed to become 'integral parts of agriculture and rural development efforts, to develop and implement energy technologies as early as possible, to require the minimum of capital investment and to be simple and inexpensive to use and maintain, to be acceptable to the people using them, and to be transferable from one region of the world to another.'[3]

Given this broad range of difficult and complex objectives, given the short space of time involved, and given diminishing Western interest in alternative energy sources, it is perhaps not surprising that bilateral and multilateral donors began to shy away from alternative energy in the late 1980s. This was unfortunate, not least because oil prices and supply simply cannot be taken for granted. The 1991 Gulf War was a vivid, expensive reminder of this. From the perspective of January 1997, when oil was selling for $25 a barrel, December 1998 – at about $9.00 – looked pretty good. But there can be no silver lining without a cloud, and in the case of oil, the cloud is its turbulent history of roller-coaster prices. In January 1999, the futures market was betting on a 14 per cent increase a year hence, a better return than anything offered by the average bank, and considerably better than cocoa at 10.5 per cent, coffee at 5.5 and cotton at minus 7.7 per cent. OPEC, in fact, generally thought to be dead until they met a few weeks later, changed the price overnight by agreeing to reduce production. The price of oil shot up within days, and the 14 per cent rise predicted in January turned out to be a 189 per cent increase by February 2000. (The projection for cotton was almost exactly on target, but cocoa fell 24 per cent during 1999, and arabica coffee dropped 16 per cent.)[4]

If there is a bite to oil dependence in the North, there is a double bite in the South. Its price and availability have indirect but significant impact on the poor through the prices of almost everything they buy. Ironically, however, few of the poor derive any direct benefit from it. Except for limited amounts of kerosene for lamps, and possibly diesel for irrigation pumps, most of the world's poor have little use for oil. For them, wood, twigs, leaves, corn husks – 'biomass' to the *cognoscenti* – is the most important form of energy after human and draught-animal power. It is perhaps more because of environmental concerns than worry about oil, that energy has remained a development concern. The story of the fuel-efficient stove is a case in point.

Stoves

In most traditional societies the hearth is the social and physical centre of the home. People cook on the fire, they keep warm by the fire; the fireplace is the focal point of the family. In fact, the Latin word for 'fireplace' is *focus*, which in the English language gradually came to mean 'point of convergence'. And in the development of fuel-efficient stoves, many important lessons about appropriate technology have found a convergence. A common element in many NGO programmes from Tierra del Fuego to the depths of the Indonesian rainforest, fuel-efficient stoves have in some cases burned more money than anything else. In others, they have become spectacular examples of successful technology diffusion.

Almost every description of a stoves project begins with trees rather than people. Made of cement, steel, iron, mud or ceramics, fuel-efficient stoves usually start their odyssey into the homes of the poor as part of an effort to reduce deforestation. The Ethiopian National Energy Commission, the Ceylon Electricity Board, GTZ/World Bank/UNDP stoves in Niger, the Ministry of Environment in Burkina Faso – all focused first and foremost on reducing fuel consumption and, therefore, the threat to dwindling forest resources. DFID's 1997 White Paper on international development said that 'new wood-burning stoves which cut the amount of fuel wood needed for cooking by half have been developed costing as little as £2. This in turn reduces deforestation and air pollution, and the cost can be recovered in just a few days.'[5]

In fact this first objective of most stoves programmes has almost always been misconceived. Successful domestic rural stoves programmes do reduce fuel consumption. But with the exception of already severely deforested areas such as the Sahelian, Andean and Himalayan regions, they rarely have a meaningful impact on deforestation. Usually long after a stove programme has failed or succeeded, it has been discovered that most rural people simply do not cut down trees for domestic firewood.[6] Certainly the average woman does not cook with green wood; neither would the average aid-agency project officer once he discovered what bad fuel it makes. A cook will use agricultural residue, twigs and branches (as opposed to trees), very often from her family's own land. Urban stoves, on the other hand, do use wood or charcoal, but this is often a byproduct of logging, and increased fuel efficiency does not usually alter the rate of deforestation.

It was secondary objectives that most often contributed to success in the diffusion of stoves. Increased fuel efficiency which is translated into reduced fuel costs, or less time in the kitchen, or fewer hours spent in collecting firewood have had the most impact. Another byproduct of many stoves programmes, one rarely mentioned in the objectives of early projects, is improved health. Acute respiratory infections are responsible for nearly a third of all under-five childhood deaths. A study of 15 developing country situations published in *The Lancet* estimated that a child exposed to regular concentration of 1600 micrograms per cubic metre of particulate concentration from biofuel combustion will have about five extra attacks of acute respiratory infections annually. In plain English this means that smoke is bad for the health, and too much smoke is very bad. And not just for children. The study also showed that in only four out of 15 cases was the smoke *sometimes* less than 1600 $\mu g/m^3$. In most cases it was more, by a factor of as much as 13.[7]

Ironically, once the message that stoves do not reduce deforestation got through to larger donors, including USAID, the World Bank and UNDP, they dropped their support for stoves programmes.[8] Saving trees had apparently been a more important reason for their investments than saving the time or health of women.

Two success stories

The Kenyan ceramic *jiko* seems to prove the exception to most rules about appropriate technology, and like other stove programmes, it succeeded almost despite the grand objectives of the original designers. It grew out of a USAID-sponsored renewable energy project with Kenya's Ministry of Energy, the sole aim of which was a reduction in the use of fuelwood.[9] Traditionally, Kenyan city dwellers cooked on a small, charcoal-fired metal stove, usually made in the informal sector from old oil drums. With assistance from ITDG, a similar stove with a ceramic liner was found in Thailand. Various adaptations were tested in Kenya, and it was found that the ceramic lining helped diffuse and retain heat, lowered carbon monoxide emissions, reduced cooking time and cut down on burns. In a laboratory setting it achieved fuel savings of 50 per cent, and in home use savings ranged upwards from 25 per cent. Training and demonstration work were carried out to a large extent by a Kenyan NGO, KENGO, with support from ATI. And like many other successful projects, the Kenyan *jiko* had a particular champion, a Kenyan named Max Kinyanjui, who added heart and soul, as well as blood, sweat and tears to the anodyne planning, research and marketing with which most reports are concerned.

Within the first four years of its introduction, 180 000 *jikos* were produced and sold in Kenya, and by 1987 when the counting stopped, it was estimated that 84 000 were being made every year. There were several reasons for the success of the *jiko*. First, although its developers had 'gone technology-shopping' in Thailand and then seemingly 'transferred' a technology they had found there, the new stove had, in fact, evolved through several prototype designs and had been rigorously tested before release. It helped that the new stove was similar in appearance to the one already in use. But it also helped that it *looked* as though it had been improved. When a genuinely improved version was introduced in Thailand, it did not sell because it did not look any different from the traditional bucket stove. The 'new improved' Kenyan ceramic *jiko* had a bell-bottom shape and was painted – two visible but technologically useless 'improvements'. The new stove, however, was appreciably better in performance than previous Kenyan versions. The ceramic liner could be made by local potters, not far from where informal sector metal-workers produced the casing. Although, or perhaps because, the stove is now produced in factories as well as backyard workshops, the price has gradually fallen, and because of the competition, quality has been maintained.

In Sri Lanka, the success of the ceramic two-pot stove came about through an unforeseen but fortuitous combination of objectives and institutional arrangements.[10] In 1979, Sarvodaya, Sri Lanka's largest NGO, made a copy of the Guatemalan *Lorena* stove. Unlike most stove producers, Sarvodaya's primary objective focused on the person who actually used the stove. The aim was first to improve the quality of women's lives by reducing the amount of time required to gather fuel and cook, and to add to their health by improving the smoky environment in which they worked. A second objective related to more hygienic food preparation. Although a part of it, forests came last.

The *Lorena* stove was not a success in Sri Lanka. Because it took longer to light, it took longer to cook a meal. It had four cooking holes, two of which were unnecessary in the average Sri Lankan kitchen, and it cost significantly more than the traditional stove. Visitors from the Indonesian NGO, Dian Desa, introduced Sarvodaya to a ceramic stove they had developed, but its design was too complex for Sri Lankan potters. It was, however, simplified into a three-part model which could be assembled in the home by a mason using straw, mud and dung to hold the components in place.

Because Sarvodaya worked in five thousand different villages, it could test hundreds of stoves under real-life conditions; the results were positive. There were problems of quality control in the manufacturing process, as some potters were better at making stoves than others, but the main difficulty related to cost. The traditional mud stove used in rural areas, being made at home, was free. Firewood was not a problem; it was relatively easy to obtain and it, too, was usually free. The trade-off, therefore, was not between the price of the stove and the price of wood. It was between having a child or a woman go in search of wood on the one hand, and purchasing a stove on the other. Even at a low price, the choice – for the average male – was obvious. Dissemination in the countryside, therefore, became a problem of *selling* the stove, almost regardless of price.

It was once again concern about trees that moved the stove to the next stage. In 1984, Sri Lanka's Ministry of Power and Energy did some calculations. It estimated that the average annual household consumption of firewood was 2.5 tons. If the Sarvodaya stove saved 20 per cent, and was placed in half a million homes, it could save 250 000 tons or 2.5 per cent of the annual energy consumption of the country.[11] Although the amount was not great, the government decided to take action, and created – as part of a national fuelwood conservation programme – an Alternative Energy Unit under the Ceylon Electricity Board. A key to the eventual success of the Sri Lankan stoves programme was the appointment by the Ceylon Electricity Board of a man with rare drive and commitment to the project, R.M. Amarasekera.

With financial assistance from the Netherlands Government and technical assistance from ITDG, the initial emphasis was placed on rural areas, and a quarter of a million Sarvodaya stoves were distributed. Because fuel in the rural areas had little cash value, however, it was necessary to subsidize the stove, making the programme non-sustainable in a strictly financial sense. In 1987 an urban programme was launched, with financial support from the UK's Department for Overseas Development (DFID; then ODA) and technical assistance from ITDG. Here the situation was very different, because the urban areas depended on firewood trucked in from the countryside. ITDG first made a detailed study of the potential market and concluded that if the stove was to be sold in any volume, the price would have to be directly related to the cost of fuel, and there would have to be a payback period of less than a month.[12] In other words, the stove would have to sell for less than the cost of a month's supply of wood if it was to find its way into the hands of the poor. Because the project aimed to become self-financing, the selling price, in turn, determined the upper limits on the cost of production.

In test situations, it had been found that poor workmanship and use of incorrect clay resulted in too many breakages — at a cost to reputation and sales. It was decided, therefore, that if quality control was to be assured, production had to be placed into a limited number of semi-skilled hands: tile- and brickmakers in urban areas and potters elsewhere. A solution was found near the town of Negombo, north of Colombo, where there are dozens, if not hundreds of tile- and brick-makers, all with kilns, and most using simple, labour-intensive technology.

Protracted and sometimes painful negotiations with tile-makers continued for two years. Experiments were undertaken to find the best clay, and to develop a process to keep the cost of the stove within the limits determined by the market survey. Finally, an arrangement was made with interested firms on four conditions: the firm had to have an existing relationship with potters, had to be willing to invest and to provide space and equipment, had to have storage space and excess capacity, and had to have adequate access to clay. The arrangement was similar to a franchise operation, for in return, the project would provide the potters with training, follow-up, and quality control. The project would also identify wholesalers, would mount a promotional campaign with potential retailers, and would conduct a general publicity and advertising campaign.

Despite intensive planning, the evolution of the project took its own course, demonstrating the limitations of blueprints. The relationship between an international NGO, the Ceylon Electricity Board and Negombo tile makers was not exactly a marriage made in heaven, nor was it one that might have been imagined when the first *Lorena* stove was produced in

1979. Nevertheless, there were remarkable achievements in a relatively short period. In July 1987, one factory with four employees turned out 480 stoves. By December 1988, seven factories were in production, employing 55 workers and producing almost 9000 stoves each month.[13]

At that rate, the annual production and sales target of 100 000 seemed more than achievable. But in 1989, sales began to fall off, settling at something between two and four thousand a month. There were several probable reasons. The market may have been saturated once the first hundred thousand were sold, and further sales reflected only replacements and a smaller number of new buyers. Quality control on early stoves was not as rigid as it should have been and complaints found their way into the market-place. This problem was compounded by a weak wholesaling arrangement and by potters outside the programme producing look-alike copies at a lower price. Imitation may be the sincerest form of flattery, but when the clay is wrong, stoves crack and sales drop. Undeterred, R.M. Amarasekera created a new organization, the Integrated Development Association (IDEA) to train potters and to deal with the wholesaling problem in semi-urban and rural areas. By 1998, at last a success, an estimated 60 000 stoves were being sold every year on a commercial basis.

If the long and winding road to this eventual success seems like an expensive process, it is worth doing a simple cost–benefit analysis. In 1998, ITDG estimated that about £400 000 had been spent on stove development over the previous 20 years. And consumers had spent about £420 000 to buy stoves. If the average householder spends Rs 90 per month on fuel and saves 33 per cent with the new stove, the total savings for only 300 000 users over ten years would have been more than Rs 1000 million.[14] Even if this calculation is four times more optimistic than warranted, the internal rate of return would still be almost 1000 per cent on project funds. And this calculation does not include jobs created, time saved, health benefits or environmental considerations.

In a field that had seen dozens, if not hundreds of failures, the Sri Lanka stoves achievement is impressive. The tide turned when the effort began to focus more on women than the environment – more on the consumer than the producer. In addition to demonstrating that it was essential to get the objectives and the technology right, it showed the importance of planning. *Before* going into major production, detailed, *professional* costing exercises were carried out, and the production scale and process were then determined on the basis of an appropriate selling price. Although this is a standard business technique, it is oddly unknown (or at least not often practised) among development agencies. As a result, many a 'good' stove has either been unnecessarily subsidized or has failed in the market-place. Or it has been purchased by those who can afford it, rather than by the poorest for whom it is intended. There were two other factors which made these two stoves programmes successful. The first was that they succeeded mostly in urban and semi-urban markets, bearing out the observation that fuel – for good reasons or bad – is not a problem for individual families in the rural areas. For this reason greater efforts in many countries are being placed on commercial, rather than domestic fuel-efficient stoves – for village baking, brewing and fish smoking. Secondly, each of the successful stoves had a *champion*; Max Kinyanjui, R.M. Amarasekera and others were individuals for whom the stove – its design, its manufacture, its dissemination – was a genuine labour of love.

Living better, electrically

There is nothing experimental about conventional electricity systems. Electricity is a 'mature' technology, one that is well known and well understood in most countries. The cost of electrification is high, however, especially in rural areas. The result is that only a fraction of rural homes in Latin America, Asia and North Africa have electricity, and in much of rural Africa electricity is virtually unknown.

Although the strongest initial demand is usually for lighting, electricity is an important infrastructure investment, especially in increasing the diffusion of development options away from urban centres. It is essential to the establishment of garages, repair shops and small-scale industries. It can be used to improve the milling of grain or the pressing of oil seeds; for crop drying, cane crushing and other food-processing equipment; for sawmills, mechanical workshops and irrigation pumps. Electric pumps in India and Bangladesh, for example, offer a considerable saving in operation over diesel pumps. Decisions about rural electrification, therefore, reflect some of today's most critical development questions: the balance between rural and urban investment; the role of the private and public sectors; the choice of technology; and the growing gap between the rich and the poor.

Oil-fired generating plants were built in huge numbers in the 1960s, but with the oil crises, attention turned to hydro-electric power generation. Choosing between hydro, where the potential exists, and coal or diesel generation is not always straightforward, however. Diesel incurs fewer capital costs, but has higher running costs. As noted in Chapter VIII, on the other hand, some of the hydro-electric projects in the South are among the most expensive and ambitious anywhere, and their environmental impact has become a major consideration in cost projections. Nuclear power, once seen as a salvation, is high in capital cost, and most projects in the South have run into a myriad of technical, financial and political problems.

Rural electrification in developing countries is often a self-conscious attempt to accelerate the pace of development and, understandably, in a different way. But modern technologies are increasingly sophisticated and costly, and developing countries start with fewer technical and economic resources, and with lower levels of rural development.

Options

Donor involvement in renewable sources of energy for power generation has been both short-sighted and inadequate. There is nothing especially sacrosanct about renewable energy unless it can out-perform the more mature approaches technically and economically over time. The cost of photovoltaic electricity was said by its proponents to have dropped between 1970 and 1990 from $30 per kilowatt-hour to only 30 cents, and it was expected to drop to one third of that by the turn of the century. Neither happened. Although investments in photovoltaic energy have been enormous, cost-effectiveness is not recognized as a 'frequently asked question' on the website of the US Department of Energy's Photovoltaics Program. Many success stories are available in the growing literature on photovoltaics – about installations at a hotel in Belize, an orangutan station in Borneo, and the Cherry Creek State Park campsite in Colorado – but price is never mentioned. The truth is that in order to generate enough electricity in cloudy Germany for an average family of four, 30 square metres of solar cells would be required, costing about $50 000 in 1998. The price would drop to $15 000, however, if the house used 48 volt DC appliances.[15] The problem is that nobody makes 48 volt DC appliances, so real-life affordable applications of photovoltaic energy have been limited mainly to solar lanterns, street lights and water pumps. In India, soft loans and tax breaks are widely available for photovoltaic installations, but the uptake continues to be low because cost-effectiveness remains problematic – even for potential middle-class consumers. And in rural areas where the technology might make a real difference if the price was right, service back-up is limited or non-existent.[16]

This is not to suggest that photovoltaic energy will never have a serious application in developing countries. The current high levels of investment are bound to yield major pay-offs in the future. The point to keep in mind about this technology – and so many others that have been crated up and shipped South in years gone by – is that price and reliability are essential ingredients of success. This is as true of a $3 ceramic stove as it is of a $50 000 photovoltaic system.

Those searching for *big* technologies and *big* breakthroughs continue to miss the importance of *small*. In 1988, 45 million small, compact, fluorescent light bulbs were sold worldwide. By 1995 the number had grown to 240 million, saving in eight years the equivalent of the electricity produced by 28 large, coal-fired power plants.[17] China has moved to the forefront in the manufacture of these bulbs for precisely that reason — to reduce the demand for coal-fired energy.

There are other alternatives to more coal, diesel and monster hydro-electric dams. One of the most elementary of appropriate technologies is the maintenance and good management of existing systems. Because of frequent power shutdowns, many Indian farmers keep spare diesel systems, an investment worth tens of millions of dollars. Power cuts accounted for a third to a half of the loss of productive capacity in the Indian cement industry in the mid-1980s.[18] During the same period, losses to Pakistan's power system through daily blackouts were as high as 38 per cent. A survey of 307 Indonesian manufacturers found that 64 per cent had their own generators because of frequent breakdowns in the public service. This was a matter of necessity rather than convenience, demonstrated by the fact that self-provided electricity cost 24 times as much as the public supply — when it was available. In Nigeria the situation has been worse for two decades. Out of 179 firms surveyed in the mid 1990s, 92 per cent had their own generators.[19] An eight-day blackout in Buenos Aires in 1999 caused an estimated loss of one billion dollars to business and industry, and resulted in a fine of $100 million to the newly privatized utility, Edusur.[20]

The windmill is not a new or experimental technology. Windmills become experimental when attempts are made to have them do new things, or when they are sited in places with weak wind regimes. Argentina has an estimated 60 000 operational windmills and accounts for a significant proportion of the world's production of small windmills. There, successful diffusion was the result of four things. The first was adequate wind for the purpose envisaged. The second was a realistic purpose — to fill water tanks for cattle. Where more ambitious purposes have been sought, such as crop irrigation or electricity generation, success in Argentina has been much more elusive. The third was the existence of a light engineering industry that permitted the manufacture and repair of windmills on a widespread basis. And a fourth was the investment of a private American firm in local production. Using designs that were already 80 years old, Aeromotor's main objective in the 1970s was to produce for export to the United States. With production costs half of what they were in the US, the result benefited both countries.[21]

Where electricity generation in developing countries is concerned, there are new technological breakthroughs on the horizon, as decades of Northern investment and experimentation in high-tech wind-generated electricity begin to pay off. Between 1998 and 1999, US wind-powered energy generation grew from 1600 megawatts to 2500, and with the cost down to 5 cents per kilowatt-hour (25 per cent less than consumers were then paying for mainstream alternatives), the government predicted that 5 per cent of the country's electricity would be produced by wind in 2020.[22]

In 1999, Denmark was generating seven per cent of its energy requirement using 4900 wind turbines, and half the country's energy needs are projected to come from wind power by 2030. Denmark dominates the $1.5 billion global wind-generator market, a nice position to be in as the price of wind power starts to become cost-competitive with nuclear, coal, oil and gas generation.

Let there be light

'Micro-hydro' refers to a power-generating plant with output of less than 100 kilowatts; 'mini-hydro' conventionally refers to plants generating between 100 kW and 500 kW, and 'small hydro' covers the range between 500 kW and 1000 kW. In industrialized countries a 100 megawatt power plant is considered small; nevertheless, micro-power units can be a cheap and well-tested technology for remote areas and small communities with appropriate water resources. Small hydro was a common means of power generation in the early years of the twentieth century, and only began to lose favour when high-voltage transmission lines were developed. In 1914, for example, there were an estimated 7000 small hydro plants in Switzerland,[23] and before the advent of national grids they were a common means of power generation in many other countries. In Sri Lanka an estimated 600 micro-power plants were installed before 1956.

With the development of national grids, however, and ever-increasing amounts of cheap fossil fuels, interest in micro-hydro diminished and existing plants fell into disrepair. Renewed interest in the technology began in China. In the fifteen years after 1970, it is estimated that China established a phenomenal 76 000 small hydro plants, generating 9500 megawatts of power, as part of what the World Bank termed 'the most massive rural electrification effort ever attempted in the developing world.'[24] The total worldwide potential for stand-alone small hydro power plants may be as high as 100 000 megawatts, of which only a fraction has so far been tapped.[25] In Peru, for example, the small hydro potential is estimated at 12 000 megawatts against a total installed capacity of only 3000. Potential of a similar magnitude exists in Thailand, Indonesia, Pakistan, Sri Lanka and Jamaica, while lesser but not insignificant potential exists in India, the Philippines, Bolivia, and a number of African countries.[26]

In Nepal there has been considerable success in introducing micro- and mini-hydro technology.[27] The story there began in the 1960s when two independent bodies, the United Missions to Nepal (UMN) and BYS, a Swiss–Nepali engineering firm, began to produce non-electric turbines for rural milling projects. In 1975, with a small grant from the Mennonite Central Committee and CIDA, the UMN began to develop a pilot hydro-electric project, followed not long afterwards by BYS. As the technology was being refined and the pilot phase ended, a need for credit arose: because of the cost, farmers and entrepreneurs would need help in establishing plants. The Agricultural Development Bank of Nepal (ADBN), which had an appropriate technology unit of its own, stepped in and provided five to ten year loans for up to 80 per cent of the cost of an approved installation. Training and extension, key elements of the programme, were provided by the manufacturers and the ADBN.

In the early 1980s the Asian Development Bank also became involved, extending a loan to the Government of Nepal in order to expand credit operations. By 1987 the number of manufacturers had risen to seven, and by 1998 there over 1200 turbines in operation, 900 producing mechanical power for agro-processing, and another 300 producing electricity. Although the benefits of the schemes are difficult to quantify precisely, some general observations are possible. They represent major savings in time, and sometimes money, over traditional techniques in the milling of wheat, maize and rice. And with some crops, such as mustard, yields have increased by as much as 30 per cent.[28] The micro-hydro schemes are significantly cheaper than large-scale hydro electric power generation. More broadly, micro- and mini-hydro schemes, like wind turbines, photovoltaic schemes and nuclear energy, produce virtually no carbon dioxide. Over 25 years, a typical 200 kW mini-hydro scheme produces about 224 tonnes of carbon dioxide, compared with 23 000 tonnes from a similar size oil-fired station.[29]

One of the drawbacks of mini-hydro is its relatively high start-up cost, but this soon balances out because of its low running costs. *La Suerte*, a Bolivian gold mining cooperative, compared its 130 kW diesel generator with a mini-hydro scheme commissioned in 1987. The installed cost of the diesel plant was 3.6 times cheaper than the mini-hydro, but because the hydro generator required no fuel and few spare parts, its annual operating cost of $22 300 was almost one third of the diesel's $59 600. With the mini-hydro loan only half paid off in 1992, it was producing electricity for 8 ¢ per kWh, compared with the diesel's 20 ¢. When the loan was fully paid, the cost was expected to fall to 1.5 ¢ per kWh.[30]

The ITDG effort in Sri Lanka provides an interesting case for comparison.[31] Grid electricity prices increased rapidly during the 1970s, and there was renewed interest in old micro-hydro schemes that had been installed during the first half of the century. Although much of the electrical equipment was irreparable, the civil works – weirs, canals and penstocks (the pipe through which the water falls to the turbine) – were in reasonably good condition.

Most Sri Lankan tea estates had such schemes, and because each had several trained engineers and a ready demand for electricity, it was thought that they would provide a good testing ground for a renewal of the technology. For the tea estates, the project was of interest because locally generated power promised to reduce costs, and because power cuts from the grid had enormous consequences. A power failure of less than four hours will lower the quality (and therefore the price) of the tea in production. Cuts of eight hours or more can result in the complete loss of the tea in process at a cost of several million rupees.

In the early 1980s, ITDG worked on the rehabilitation of several schemes in the tea estates, one of which had originally been installed in 1898. Working relationships were established with a small British firm, Evans Engineering, for the production of new turbines, and with a Sri Lankan firm, Brown and Company, on their installation. One of the most important technological developments was an electronic load controller. It is essential that the frequency of electricity remain constant. But fluctuations in demand require a governor which can alter the water flow in relation to the changing demand. This presents a problem which, in the past, was controlled mechanically with centrifugal governors linked to the water valves. Oil hydraulic systems are used on most modern turbines, but for small schemes they have serious drawbacks: they are expensive, difficult to repair, and they require a relatively sophisticated degree of on-going care and maintenance.

Various experimental electronic-load controllers (ELC) were under development in Europe at the time, but none suited the Sri Lankan requirements. With ITDG funding and technical involvement from Evans Engineering, however, a suitable prototype was produced by another small British firm, GP Electronics, and by the late 1980s, over 100 of their units had been exported and successfully installed in China, Thailand, Malaysia and other countries. The ELC made a significant difference to the cost of renovating the Sri Lankan micro-hydro plants, to other ITDG micro-hydro projects in Colombia and Peru, and to some of the Nepal projects.

By 1983, ITDG had refined the technology to the point where, with good training and follow-up, it was replicable and sustainable. But there were worries within the organization that simply rehabilitating micro-hydro projects on nationalized tea estates did little for the poor. Reducing the cost of tea production would add to the country's foreign exchange earnings and might improve the balance of trade, but the workers on the estates themselves – descendants of low-caste Tamils who had migrated from India in the nineteenth century – remained among the poorest in the country. A unique scheme was therefore developed in conjunction with a Dutch-funded Integrated Rural Development Programme (IRDP). The IRDP would fund 30 new schemes on the condition that the net surplus produced through the reduction in use of grid electricity was paid into a development fund for estate workers. By 1998, there were 40 village hydro schemes, each locally owned and managed, and ITDG had become the *de facto* national reference on micro-hydro.

Dances with wolves

Institutional donors clamour for 'lessons learned'. In this story, unfortunately, several donors proved unwilling or unable to learn. Having demonstrated the effectiveness of the technology by the mid-1980s, and having gained the confidence of government and the tea estates, ITDG expected large donor agencies to come forward to support the rehabilitation of some of the hundreds of schemes that remained untouched. They did, but not in the way that might have been anticipated: the perceived opportunity was for sales, not power generation. The first to arrive, in 1983, were the Chinese. Engaged by the World Bank to write specifications for ten pilot schemes, Chinese engineers not surprisingly drew up specifications that could be met only by Chinese equipment. Although several international firms tendered for the contract, the bulk of it went to China. The results were not good. The Chinese equipment looked impressive, but without follow-up and spares, without sufficient training for the operators, without a Sri Lankan agent and local technicians to provide technical assistance, one by one the systems failed, mainly because of problems with their complex hydraulic governors.

Then came the Canadians. Under CIDA auspices, a consulting firm produced an extensive report which, among other things, projected an internal rate of return that was so high (and so wrong), that an overnight fortune seemed possible in micro-hydro. The report, which cost almost as much as ITDG had spent on micro-hydro in Sri Lanka altogether, was quietly buried. Then, in 1985, the British began to take an interest. Extrapolating from the ITDG experience and projecting a possible budget in the neighbourhood of £10 million, DFID thought it might be possible to renovate 140 or more schemes.

A British consulting firm was engaged to prepare feasibility studies, but things gradually turned sour. The DFID objectives had little to do with appropriate technology, local production or cost minimization. Knowing this, the consulting firm designed systems that maximized British content: high-cost turbines rather than the basic versions used by ITDG; British switchboards rather than the perfectly adequate varieties that had been made in Sri Lanka for half a century; concrete-lined ductile iron penstock – the Rolls-Royce of penstocks. Apart from the effort to maximize British content, there was no particular reason for the firm to specify anything less than the very best, for if the project went ahead, they would have minimized risk and confusion (and possible damage to their reputation) by avoiding unknown products, small firms and Sri Lankan manufacturers.

Sri Lankan engineers, unimpressed with the technical quality of the reports, were shocked to discover that British engineers, with no direct experience of micro-hydro, were to be charged at *more than one hundred times the salary of a Sri Lankan engineer*. Negotiations became more strained when it was proposed that DFID would supply the equipment and consultancies (the consultancies alone would take up 25 per cent of the budget) and Sri Lanka would cover the cost of the civil works. Half of the £10 million, therefore, would be provided by the Government of Sri Lanka, but the entire amount would be charged to the tea estates as a *loan*.

This made it imperative to the tea estates that the project be as cost-effective as possible, which it clearly was not. From a minimum of 140 schemes, the proposal changed to a maximum of 60. Suddenly an appropriate technology no longer looked appropriate or even feasible. Worse, in anticipation of new schemes, some of the estates had run down their spare parts supplies and had neglected preventive maintenance, leading to costly damage and closure of some existing plants. By 1989, the ODA scheme had dissolved into a modest, face-saving proposal for two pilot schemes.

Lessons

If there is a single lesson to be drawn out of energy projects, it is the importance of setting clear and reasonable objectives. Success comes from dealing with people's immediate needs rather than with vague concerns about the environment. (Einstein once defined environment as 'everything that isn't me'.) As important, perhaps, using poor people and poor countries as guinea pigs for underdeveloped technologies is likely to result in failures which simply reduce future options. In terms of reaching the poor, the most successful examples in this chapter are the Sri Lankan and Kenyan stoves, both of which were similar to known, traditional stoves, both of which were cheap and which provided clear, short payback periods.

Several of the projects were advanced by the existence of a local manufacturing base which could produce and repair components. And the successful projects all benefited from systematic monitoring and evaluation. Gradual evolution and adaptation were the rule in most cases, and this required time. Short, unrealistic time-frames are the cause of many failures that are often unfairly ascribed to the technology. One of the reasons for the 'success' of the Chinese biogas and micro-hydro programmes, for example, was the huge government investment in research, training, dissemination and follow-up over a twenty-year period. This was accompanied by a willingness to tolerate failure and to learn from it. An estimated 85 per cent of the Chinese biogas plants installed between 1973 and 1978 – probably more than have ever been produced in all other countries combined –

failed not long after installation. Andrew Barnett correctly warns about concepts of success and failure: 'In industrialized countries it is common to think of technologies taking more than 30 years from the initial innovation to total *saturation* of the market. In many cases, the investment in energy technologies in the rural Third World has been so modest and so recent that questions of success are premature.'[32]

In virtually all of the cases described, the commercial or semi-commercial sectors played a vital role in the ultimate dissemination of the technology – providing credit in the case of Nepal's micro-hydro; investment in the case of the Argentine windmills; technical expertise in the case of Sri Lanka micro-hydro; manufacturing know-how and marketing expertise in the case of the Sri Lankan stove. The role of government was also important in most cases, in recognizing – whether in connection with Sri Lankan stoves or Nepalese micro-hydro – that alternatives to simply 'extending the grid' are essential if the rural poor are to be reached within the lifetime of those already born.

And in each of the cases discussed, development agencies played a key role. NGOs and individuals with a basic interest in appropriate or alternative forms of energy were often important to a success. Interestingly, most of the cases also involved one or more bilateral agencies. In most instances, except that of micro-hydro in Sri Lanka, where things went badly wrong, this involvement was constructive and supportive.

CHAPTER X

The house that Jack built: construction materials

I tell this tale, which is strictly true,
Just by way of convincing you
How very little, since things were made,
Things have altered in the building trade.

Rudyard Kipling, *A Truthful Song*

ONE BILLION, THREE hundred million people live in absolute poverty. They, and probably a lot more, have bad diets, poor health and little education. It goes without saying that they also lack adequate shelter and live in unsanitary, unsafe conditions that provide little protection from the elements. There is no lack of good ideas and practical solutions for the problem. It is the *availability* of affordable, durable construction materials that determines how much and what type of construction will take place in a given country. Expenditure on building materials in the South absorbs between three and five per cent of GDP and represents an immense drain on foreign exchange. Much of the bill is charged to Africa, where in some countries over 90 per cent of the value of commercial building materials is imported.[1] The real challenge, therefore, is to ensure that affordable, local, durable materials are widely enough produced and in sufficient quantities that they can reach the poor and, where possible, create jobs in the process.

Much of what is known about building construction has been known for centuries, and in many countries of the South, the old ways remain the best available for most people. But with increasing populations and growing energy costs, the old ways are no longer always efficient. In some cases they have been superseded by modern technologies which produce the same materials – bricks, mortar, roofing materials – at a high cost in capital equipment, and at a loss in employment opportunities.

This chapter examines four intermediate technologies in construction materials: lime production in Malawi, cement making in India, the production of bricks in Botswana, and the development of a new building product that looks like an old one – fibre-reinforced concrete roofing tiles. In addition to the lessons these cases have to teach about construction materials, they also carry messages about diffusion, and about the thorny question of profit: what it is, how it is calculated, and what happens when it is suppressed.

The Balaka lime project

Malawi, a beautiful and sometimes forgotten central African country, is also one of the poorest in the world. Only nine others have a lower per capita GNP; at $220 in 1997, Malawi's was 20 per cent lower than that of Bangladesh. Roughly five million Malawians, out of a total population of ten million, are of working age, up from 3.7 million only a decade before. Population growth will continue to push this number up, but employment for job-seekers is in short supply.

Starting around 1980, several institutions and programmes were established in Malawi to assist small- and medium-scale enterprises, with an emphasis on rural, off-farm job creation and import substitution. One of these institutions, Indefund, supported a pilot scheme for the production of hydrated lime. Lime production for construction purposes dates back to Ancient Egypt, and lime producers using old techniques can still be found throughout Latin America, Asia and Africa. Limestone, plentiful in the Chenkumbi Hills of south-central Malawi, is mined by traditional lime producers, but their yield is low and the quality poor. As a result, when the project was first approved, Malawi was importing almost half its annual requirement – 3000 tons, at a cost of approximately $325 000.[2] Roughly half the lime went to the construction industry and the rest was used as a reagent in the production of sugar. A 1987 ITDG feasibility study found that by upgrading traditional production techniques, better quality lime could be produced locally, reducing dependence on imports and creating jobs for Malawians.

The project that emerged – for the establishment of a pilot plant to produce consistent, high-quality, chemical-grade lime for the sugar and construction industries – had a secondary goal. By integrating parties with different interests and skills in the project, it was hoped that a model for further collaborative efforts might be developed. The project involved credit from Indefund, technical assistance from ITDG, and the production of kiln and factory equipment by a local engineering firm. The borrowers were four Malawian entrepreneurs, operating a small company called United Mining Company Ltd. Sited near limestone quarries at Balaka, the project involved the manufacture of a steel-encased vertical shaft kiln in which the lime would be fired. It was a continuous-feed process that required the development of a number of pieces of moderately sophisticated equipment, including a motor-driven hydrator which mixed quicklime with water after the firing process.

As with many, if not most projects of this sort, things went wrong throughout the trials. Some of the specifications had been drawn up incorrectly, and some of the manufacturing was sub-standard. A decision to switch the fuel source from charcoal to coal led to damage on the first firing, and this in turn caused delays until the kiln could be repaired and the process corrected. With the technical problems solved by about 1992, problems of ownership and management reared their unpretty heads. In an effort to created a worker-owned commercial operation and to provide the necessary investment capital, ITDG established a commercial firm, Chenkumbi Limeworks Ltd, which broke even in its second year and made its first profit in the third. By that point, the project had created employment for 90 people and had brought a significant amount of new wages into the area. On its own by 1997, however, the company failed to meet chemical-grade standards, losing its key market in the sugar industry. Added to this, absentee management and a corrupt politician dealt the company a fatal blow.

Among the key lessons contained in a 1991 evaluation were the need for good co-ordination and leadership, the importance of adequate time for the project to mature, and an ability to tolerate and learn from failure – all consistent with other successful projects focusing on new technology. A 1994 evaluation still viewed the project as a success, but urged a restructuring of the ownership and management.[3] By 1997, however, when the Minister of Works – who had a personal interest in lime production – was being taken to court over a diversion of project funds, the effort was more or less doomed.[4] The project taught its supporters useful technical lessons, and these are the ones most prominent in the literature. There are additional lessons, already observed in Zimbabwe's Murombedzi oil mill project (Chapter VIII) and in others to come. One has to do with the difficulties in creating enterprises that are beyond the investment and managerial capacities of those who will take over when the outsiders are gone. Another has to do with allowing an eagerness for technology to get in the way of management acumen and probity.

And a third is that you win some and you lose some. Dennis Rondinelli has written extensively about the unavoidable risk factor in development projects, observing that 'in most cases it is difficult or impossible to define goals and objectives precisely at the outset, or to give more than general indications of what can be accomplished when a proposal is initially made ... Rationalistic planning and management procedures often require information and data that are simply not available in most developing countries.'[5]

Mini-cement in India

In nineteenth-century Europe, cement was manufactured in vertical shaft kilns (VSK) very similar to the lime kiln developed in Malawi. Gradually, however, production shifted to a rotary kiln process which allowed continual operation and which, along with greater fuel efficiency, produced better, more uniform cement. Although it was revived and widely used in Germany during the Second World War, the VSK process has now all but disappeared from Europe. It can still be found in Australia, however, and in the 1960s began to find favour in China. In 1975, 28.3 million tons of cement, 60 per cent of the total Chinese production, was being manufactured in VSK mini-cement plants, each with a daily capacity of under 200 tons. By 1983, 81 million tons, which represented 75 per cent of the entire Chinese output, were being produced in mini-cement plants.[6]

In India, various experiments with the VSK began after the Second World War. There were basically three strands to the research. The first had a military purpose. Following India's brief war with China in 1962, the Indian Army sought the means to produce small batches of cement where and when it was needed, instead of relying on transportation from distant factories. The second research stream was commercial, inspired in part by a fit of pique. A Jodhpur flour and rice miller, D.P. Saboo, required cement for alterations to his factory, but his application for an allocation – cement being heavily rationed in the 1970s because of the oil crisis – was refused. Annoyed, Saboo decided to make his own cement. His first experiment, in his own garden, was successful enough to encourage him to continue with the experiment.

The third research strand had more conventional origins. The Cement Research Institute of India, using the military research as a starting point, began VSK experiments in the late 1960s. In addition, under the auspices of the Uttar Pradesh Government, a researcher named M.K. Garg undertook several vertical shaft kiln experiments based on efforts that the military had by then abandoned. When the government lost interest in 1972, Garg took his passion with him to the newly formed Appropriate Technology Development Association (ATDA) in Lucknow, where the experiments continued – with funding and technical support from ITDG and Appropriate Technology International.

The first impetus for Indian research into the viability of mini-cement plants may have been military in origin or simply indignation on the part of D.P. Saboo. But there were good developmental and economic reasons to undertake the research. Unlike large plants, the smaller units could be established where there were limited deposits of suitable raw material. The smaller plants could also be set up in areas where limited demand would not justify a larger plant. By producing close to the market, costs could be reduced because of shorter transportation distances.

Between 1981 when the first mini-cement plant (MCP) went into operation and 1986, some 60 commercial plants were commissioned. The mini-cement plants were, therefore, a modest success but, as with so many things in India, the success emerged from a complex web of factors not entirely planned at the outset. Part of the impetus for research had been the great shortage of cement through the 1960s and 1970s, and the strict pricing and rationing control on large plants. This climate of control may actually have discouraged production and expansion in the modern sector, and may thus have boosted the attractiveness of small-scale production. Although cement was partially 'decontrolled' in 1982, mini-cement plants remained totally exempt, which meant that they could sell at whatever price the market would bear. Where there were shortages, this gave mini-cement a modest price advantage of perhaps five or ten per cent over the larger producers.[7]

When a mini-cement plant is compared with the more standard large-scale variety, the actual cost of producing a ton of cement is almost the same.[8] But when average market prices are compared, the advantage enjoyed by the smaller plant in being located closer to its customers gives it a further advantage. As a percentage of total price, transportation accounts for as much as 30 per cent in the case of large-scale production, and only 10 to 11 per cent in the case of mini-cement. Beyond the price advantage, in a market where adulteration is common, local production has the advantage of more intimate customer–client relationship; trust, therefore, is also a factor.

The major limitation of the MCP has been its less efficient use of energy. Early Indian mini-cement plants used between 11 and 22 per cent more power per unit of production than the larger plants, although newer models claim greater efficiencies. The MCP has several advantages besides its ability to produce a price-competitive, high-quality product. On average, the MCP requires 25 per cent less capital investment per unit of output than the large plants. And unlike the rotary kiln plants, which have a foreign exchange component of as much as 15 per cent, the VSK has none. The job creation factor is also important. In addition to being more labour-intensive, an MCP requires fewer highly skilled workers than a large plant.

MCP technology is not for the micro-entrepreneur. India's National Development Research Corporation was selling a 100 ton per day plant in 1999 for $3.9 million, but even at that price, it was a fraction of the cost of large-scale technology. Mini-cement should not be expected to replace large-scale production, but it does have a solid place in the technology of cement production and represents an economically beneficial alternative.

It is worth noting that the mini-cement plants and the costings used in this chapter all relate to standard Portland cement of the same quality as produced in the large-scale plants. Portland cement is the minimum standard which can be sold in India, but in China there are six permissible grades. While Portland cement is specified for bridges and multi-storey buildings, lower grades suffice for single-storey dwellings and other less-demanding uses.[9] The Chinese policy had a major and positive effect on the diffusion of mini-cement plants in that country, many of which do not have to produce to the high specifications demanded of their Indian counterparts. Indian mini-cement plants thus came into being *despite* the lack of an appropriate government policy on cement grades.

The policy issue is one strand of research on a different approach to cement undertaken by the International Development Research Centre (IDRC). Going back to basics, IDRC studied the properties of cement used in Egyptian pyramids and in Roman construction. Known as 'pozzolan' or 'pozzolana' cement, it was named after Pozzuoli, a town in a volcanic part of south-western Italy. Early builders had discovered that pozzolan, basically a volcanic ash with a high silica content, when ground together with lime and mixed with water, formed a durable cement. A similar mortar produced from lime and crushed brick has been produced in India for centuries, and it has been found in the ruins of Mohenjodaro in Pakistan, dating back to the third millennium BC. The modern relevance of pozzolana cement is that the production process requires significantly less energy than Portland cement. And it is based on a raw material – volcanic ash – that is widespread in many developing countries. Studies in Bolivia and Guatemala, for example, have shown that pozzolana cement could be produced at half the cost of Portland cement. IDRC research in the early 1990s identified, *inter alia*, two key issues related to the material's potential in Latin America: the need for a reliable, speedy test to evaluate the potential of pozzolan deposits, and the establishment of appropriate specifications for different types of construction.[10]

Although the economics of pozzolana cement seem straightforward, the question of attitudes is critical. As with many traditional technologies, pozzolana cement did not enjoy the benefits of modern research and marketing that were applied to Portland cement during the twentieth century, and so it remained a 'second class' technology in most of the places where it survived. In their day, the Stanley Steamer and the Studebaker Electric Automobile were highly competitive with cars powered by internal combustion engines. But – in part because of cheap oil – research, development and marketing went one way, and they went the other. This is one of the key lessons in the Indian mini-cement story: the economics of manufacturing – in almost any sphere – are situational and can change. Mini-cement production in India may have flourished despite a lack of government policy on differing cement qualities, but it was made competitive, in part, because of government restrictions on large-scale manufacturers. Had these restrictions not been present as the technology was developing, the economic advantages accruing to large-scale producers in access to credit, subsidized power and other resources might very well have made the small-scale effort less possible and less attractive, despite its obvious employment-creation advantage.

Bricks and brickbats

With a population roughly one tenth of one per cent of India's, and a completely different resource base, Botswana's approach to development necessarily differs greatly from India's. Nevertheless, there are parallels or potential parallels in the introduction and diffusion of new technologies.

In Botswana in the late 1980s, all cement had to be imported, and because local brick production was of a low quality, bricks had also to be imported.[11] Trucking bricks in from South Africa was an expensive and arduous proposition, and so most building was done with cement blocks. Although building with cement blocks was 60 per cent more expensive than with bricks, the simple unavailability of good bricks gave the advantage to cement. It also provided an opportunity for local brick production.

Brickmaking can be an entirely manual operation – from the digging of clay to the mixing, moulding and firing of the bricks. Because it is an old technology, various levels of mechanization are available, depending on scale and the desired level of investment. Machinery includes mechanized mixers, various kinds of mechanical or automatic presses and extruders, and different types of kiln. Using data from existing small-scale pressing plants and figures from a feasibility study for a large-scale plant, Raphael Kaplinsky observed much greater potential economies in small-scale Botswana brickmaking than existed in the case of Indian cement. A small-scale plant, for example, employing 29 people and turning out a million bricks annually, compares favourably in job creation with a large-scale plant employing only 4.5 times the number of workers but with 24 times the production. When capital costs are factored in, the comparison is more striking: 'Botswana's annual consumption of bricks . . . could either be met by a single plant costing P5.33 million, employing 131 people and operating at about 40 per cent of a single-shift capacity . . . or with 11 small-scale pressing plants employing 319 workers and costing P1.57 million.'[12]

The striking economy of the small-scale plant can be seen in the selling price required in order to break even over 20 years. The small-scale brick press with a purpose-built kiln could sell bricks at 43 per cent of the large-scale plant's price, and 20 per cent of the imported brick price at the time the studies were done.

For a potential investor with a calculator, such observations in the mid-1980s made brickmaking an appealing investment opportunity. In addition to traditional brickmakers, there were two such investors. The first was a Motswana who had worked in a Zimbabwean brick factory. In 1982 he had purchased used brickmaking equipment in Zimbabwe and set up a factory in Francistown. The machinery had the capacity to produce half of Botswana's annual demand at the time, but in addition to unanticipated problems with the local clay, the plant proved to be technically and managerially beyond the owner's capacity. Eventually he sold it, and in the first two years of operation, the new owners produced bricks at less than four per cent of capacity.

The second investor was an NGO, whose factory ran consistently at a loss because, Kaplinsky says, 'they were not aware of their production costs'. They sold their bricks at about 30 per cent less than actual cost, and at only 23 per cent of the cost of the imported bricks against which they were competing. Understandably, the bricks sold well but the operation required frequent injections of donor support to stay in operation.

Unlike the Balaka Lime Kiln, the partnership potential in Botswana was limited. The private investors received no technical or managerial assistance, while the NGO observed none of the financial discipline inherent in a market-based operation. The level of expertise and management skill available among potential investors was insufficient to kick-start a brickmaking industry despite its very real profit-making potential. This had left the field to NGOs which could offset continued losses with injections of cash for new 'phases' of the 'project'. Worse, by selling below cost, the NGO may actually have discouraged other potential investors, thereby actually holding back diffusion of the technology.

ITDG's experience in Zimbabwe was different. Unlike Botswana, there was already an existing low-cost, low-quality traditional approach to brickmaking, and a high-cost, high-quality modern technology, and nothing in between. Following independence in 1980, there had been a major house construction boom in urban and suburban areas, and by 1987 when ITDG became involved, the demand for bricks far exceeded the supply available from modern factories. The typical 'farm brick', on the other hand, did not meet building codes.

The traditional technology is not unlike that used elsewhere. Moulded clay bricks are dried in the sun and then stacked into a 'clamp' – a rectangular arrangement of twenty or thirty thousand bricks with firing tunnels underneath. The clamp is insulated with mud, and fires are lighted for a day or so. Although this is a well-known technology, the standard Zimbabwean product had limitations: many of the bricks were weak, misshapen and had a propensity to high water absorption. Most problematic, however, was the use of wood to fire the clamp in a country with serious deforestation and soil erosion.

The ITDG approach aimed to improve the existing technology, rather than to transform it. The first element was improved brick production using a simple table mould produced in Britain by J.P.M. Parry & Associates. The second and most important innovation was the replacement of wood with coal which is mined and readily available in Zimbabwe. Unlike other technologies described in this book, and unlike the Botswana example, there was no difficulty in disseminating the technology or in selling the bricks. By 1996, more than 60 separate businesses were using the coal-fired clamp, most in semi-urban areas where the market was greatest and where the price was affordable. Additional work on clay testing, brick moulding and the substitution of free boiler waste for coal would improve the product further, but essentially, the 'better mousetrap' had already been created and sold.

The implications of greater energy efficiency – a 25 per cent improvement in Zimbabwe – could be more profound elsewhere. Experiments in Peru, where there are severe fuelwood shortages, have shown that energy consumption can be almost halved with coal. In Sudan, where the brick industry consumes half of the country's fuelwood, the use of oil, bagasse and cow dung have proven cost-effective in pilot projects. The long-term potential in cost saving, energy reduction and job creation is even greater in Bangladesh, where brick production accounts for over 200 000 jobs, and in Sri Lanka where 85 per cent of brick production is in the hands of small and medium enterprises.[13]

Fibre concrete roofing: lessons from the Black Country

The story of fibre concrete roofing (FCR) is fascinating in both its simplicity and its complexity.[14] The technology seems simple and straightforward. Natural fibres such as coir, sisal or jute are mixed with concrete and formed into sheets with a thickness of about a centimetre. In the early stages of the product's development, corrugated roofing sheets were the main use, although these gradually gave way to the production of smaller pantiles, an S-shaped roofing tile that can be laid so the downward curve of one tile overlaps with the upward curve of the next.

Because the technology was relatively simple, because the capital investment required to set up a small plant was low, and because there is an almost insatiable demand in most developing countries for inexpensive, durable roofing material, the FCR technology, once adequately tested and developed, began to take off at great speed. By the early 1990s, FCR roofing equipment could be found in 60 countries, and as many as 500 new low-cost, labour-intensive manufacturing units were being sold every year. Both the technology and its diffusion were the product of a unique marriage between NGOs and private sector firms in Britain and developing countries. Judged on the basis of Schumacher's criteria – small, simple, cheap and non-violent – the technology is a success. Judged by the more complex criteria of some appropriate technology organizations, it also stands up well. It can meet the roofing needs of a wide majority of the community, and is 'ownable, controllable, operable and maintainable within the community it serves.'

If this book demonstrates anything, however, it is that while an appropriate technology may be small, simple, cheap and non-violent, its development is often the opposite. The efforts involved in the development of FCR were not small. Getting the technology right was far from simple, and although the research and development investments were perhaps small by the standards of construction materials giants, they were large for the players involved. And if non-violence suggests harmony among colleagues, FCR was anything but non-violent. More than any other technology described in this book, FCR succeeded in a climate of ideological acrimony, and varying degrees of stubbornness, myopia and confusion.

The story began during the early 1970s when a British building materials consultant named John Parry visited Ghana to look into brickmaking potential. Parry realized that there were easier ways of making bricks in Ghana than those favoured by his employers and, gradually falling under the influence of Fritz Schumacher, he eventually set up his own consulting firm to advise clients on more appropriate building technologies. Serving on the Board of Trustees of ITDG, he became involved in various building materials projects, experimenting first with brick-moulding equipment and then with roofing material.

By 1982, some of the new techniques and machinery had been widely tested, but Parry was frustrated that much of the equipment sent to Africa and Asia lay idle. He was especially concerned about fibre-cement roofing sheets and tiles. Prototype equipment was not replicated, materials were not being produced in any significant volume, and much of what was being produced was not being properly used. Roofing sheets cracked and broke, tiles did not fit together properly, and roofs leaked. In short, it was beginning to look like one of those not-quite appropriate technologies – a good idea that did not really work, except under tightly controlled experimental conditions.

Believing that the technology *could* work, and frustrated at 'too much running around' and not enough emphasis on the production of 'damned good equipment', Parry decided to alter the nature of his consulting firm, and go into the production of low-cost, appropriate construction-materials equipment.[15] His company, J.P.M. Parry & Associates, established a small manufacturing operation at Cradley Heath, in the Black Country near Birmingham, and Parry called it 'Intermediate Technology Workshops'.

Up to that point, several NGO attempts to programme the sheet- and tile-making equipment through small groups of women or unemployed youth had only limited success. NGOs with school building programmes, for example, would buy the equipment so that parents could make the tiles themselves and then reap the benefits of a productive enterprise afterwards. Where training was good and production was adequately controlled, high-quality school roofs followed. But few viable enterprises continued after the initial impetus. In Kenya, for example, ActionAid established several such organizations, but most survived only because ActionAid continued to purchase their output for ongoing projects elsewhere. Of ten groups in Kenya established by ITDG in the mid-1980s, only four had survived by 1990, none of them thriving.

Quality control was sometimes a problem and this led to marketing difficulties. But there were other marketing problems. Selling a new tile in a climate of economic and attitudinal competition from corrugated aluminium and galvanized iron sheets, as well as concrete and burned clay tiles, was not an easy matter for tiny entrepreneurs without experience in marketing. Some of the enterprises had been established in rural areas where markets were limited. Corruption was an additional problem in some cases. Inexperienced in bribery, some found it inexplicable that a builder would deliberately choose an inferior, more expensive product over theirs.

John Parry was undaunted. By 1985 he was focusing most of his roofing business on the equipment for making tiles. Originally an entirely manual operation, Parry developed a simple vibrating tray that produced a tile 25 cm wide and 50 cm long. The vibrator consolidated the mixture of sand, cement and fibre on a plastic sheet, de-aired the mixture, and produced a uniform thickness of either 6 or 8 millimetres. The cement, formed on a plastic sheet, was then laid on a mould in the shape of the desired tile. After drying and curing, the tile was ready for transfer to the building site.

In travelling to meet customers, Parry and his staff learned several important things that helped improve the technology. The first was that quality control in the tile making was essential. Poor mixing, poor curing, or carelessness in the vibration process could all lead to tile breakage and failure. Reducing such failures to a minimum was essential in getting the product into the market, and good training and follow-up were therefore essential. The tile requires a different roofing structure from that used with other materials; it requires more timber than galvanized iron sheet, but less than concrete or clay tiles. It was therefore important that roofers also be properly trained in installation.

The original tile-making machines were electric and could be operated from a car battery or from a mains supply. But many of the building sites where FCR might be used had neither, so a hand-crank model was developed. This was a useful alternative where labour was cheap and plentiful, but in some cases this additional cost could reduce the economic viability of the tile. So in 1990, a gyro-drive vibrator was developed which enabled one worker to handle the entire process. Basically the operator cranks a flywheel and then flicks a switch to engage the vibration. Like a child's toy car, the momentum in the flywheel continues to supply power while the operator prepares the tile. Because Parry had developed other products that used the vibration technique, including floor tiles, hollow concrete blocks and pipes, the next innovation was a detachable electric vibrator which could be used in other products as well as tiles. And for countries where fibre was either unavailable or added significantly to the cost, an alternative was found in chipped stone.

Although much of his equipment was still sold to or through development organizations, Parry had no difficulty in finding small entrepreneurs who could turn it into a successful business. The company established agents in a dozen countries and provided them with training. They in turn provided training to their customers, and in some cases sent them to Cradley Heath. Because there is generally a greater mark-up in roofing than in tile production, Parry found that some of his most successful customers were those already in the roof construction business. The tile gave them a product that was both superior to what they had been buying, and was in many countries significantly cheaper.

In 1985 the company sold about 50 sets of tile-making equipment, and by 1990, output had risen to 500. By then the company had exported equipment to 60 countries, and had steady orders from about 20, from Vietnam to Uganda to Nicaragua. By the mid-1980s, however, Parry and the NGOs with whom he had once formed a mutual admiration society had parted company in acrimony and accusation. Parry had grown increasingly disenchanted with what he saw as the amateurish NGO approach to production and marketing. Establishing producers drawn from villages and the ranks of unemployed youth had not worked, and the blame was being placed unfairly on the technology.

In refining and extending the definition of 'appropriate', on the other hand, some NGOs believed that the manufacture of tile-making equipment should be transferred to developing countries. ITDG's support for the development of the technology in the 1980s was believed by many to have been predicated on its eventual manufacture in Africa and Asia, not in Cradley Heath. If there were profits to be made in the sale of tile-making equipment, they should be generated in the South, not the North. A set of Parry tile-making equipment could be made for significantly less in developing countries than in Britain, and some set out to prove it. ActionAid, for example, helped a Kenyan NGO establish a workshop where copies of the basic equipment were made at half the Parry price and without sizeable foreign exchange considerations. In response to the obvious criticism that the Kenyan machine and other copies did not reflect the cost of research and development invested by Parry, the NGOs said that much of Parry's initial investment and travel were subsidized by development agencies. They also argued that a large proportion of his customers were still development organizations.

Parry's reaction was a broadside against shortsightedness. Indeed, the selling price of cheap imitations does not reflect the cost of research and development, and when supported by NGOs, probably does not reflect the real capital cost of manufacturing the machinery either. As in the case of brick manufacture in Botswana, tile sales themselves may not even cover the cost of production. He was less acerbic about private sector copies, citing a case in Accra. The private sector copier there had as much an interest in seeing the product function properly as Parry did; future sales depended on it. But NGOs which move in a subsidized world without a bottom line, Parry observed, going from fad to fad, have no vested interest in the success of the technology, or of the enterprises they purport to 'establish'. Often they fail to provide adequate technical advice, follow-up and repair services. Many do not know the first thing about business. They tended to be hostile towards Parry because he was creating a dependence on British goods and services without realizing that he was equally dependent on his customers in the South: dependence, he argued, is a two-way street.

Although there is cant on both sides of the debate, the scales tip in Parry's favour. A small enterprise named 'Shelter Works' in Nairobi exemplifies some of the challenges that face the small tile-maker. Shelter Works was founded in 1988 on a plot of rented land without electricity or water. The proprietor started in a shed with one battery-powered tile-making machine and 100 moulds, but was uncertain about his prospects and did not leave his full-time employment until early 1990. In the intervening period, business was touch and go. Realizing that his output was insufficient to cover even the most basic costs, he sold his car and invested in three more machines and another 750 moulds. He arranged for electricity and a water connection, and because marketing rather than production emerged as the main problem, he had a telephone installed. He also realized that potential customers for roofing tiles often needed other construction materials, so he began to diversify into concrete and stabilized soil blocks. Without allied construction materials, potential customers would sometimes visit and then go elsewhere; they wanted a one-stop dealer. By 1990, Shelter Works was a going concern with nine employees – a number that fluctuated with orders – and the proprietor was actually spending most of his time on marketing.

Shelter Works worked because the proprietor produced good tiles. This was, in part, because of his own background in mechanical engineering and because he had a foreman and workers he could rely upon while he was doing the all-important selling job. Under-capitalized at the outset, he had enough savings, in the form of his car, to expand and diversify when the need became apparent. His sales improved with the publicity attached to a large housing project that began at about the same time he went into production. The first phase of the Koma Rock Housing project used FCR tiles on 1700 middle-income houses, giving a major fillip to the product. A further benefit arose from its being certified by the Kenya Bureau of Standards. Although NGOs had sought this approval for several years, it came as much because two employees of the Bureau – listening to the sales pitch from the NGOs – bought tile-making equipment for themselves and had a personal interest in seeing it approved.

As in the case of the Tinytech oil press in Zimbabwe, the drive and skill necessary to operate a productive enterprise profitably are not things that can be plucked out of a catalogue as easily as a tile-making machine. It makes little difference where the machine is made if it does not produce, but Parry was able to do something that NGOs have difficulty with. He was able to get the machines into the hands of genuine *entrepreneurs*. Some may have been well off, but most were not. Like the proprietor of Shelter Works, or the Ghana Airways stewardess who took time between flights to go to Cradley Heath for training before buying Parry equipment, many were entrepreneurs on their first outing, individuals who wanted to augment their income or be their own boss and who, in the process, would create jobs and products for others.

At £1300 for a basic starter set of tile-making equipment in 1999, the price charged by J.P.M. Parry & Associates was still far from exorbitant. No matter what the mark-up, it was not producing massive profits for the company. In fact the Parry workshop at Cradley Heath is much more reminiscent of an NGO operation than a modern manufacturing plant. It is less well equipped than the GRATIS Workshop in Ghana or the Rural Industries Innovation Centre in Botswana, and it is half the size. Because it is located in an industrial area reminiscent of the 'Third Italy', much of the manufacture of components and test items can be farmed out to small firms in the immediate area. So a large workforce and plant are not necessary. Its staff have included former VSO volunteers, and most of its Board of Directors have international experience, several with NGOs. One shareholder and a former board member, David Astor, was an early supporter of Schumacher, and was central to the foundation of ITDG in the 1960s.

The company produces an occasional newsletter that is reminiscent of NGO newsletters, and there is a ten-point description of what it takes to be a good agent for Parry equipment. The major points could well serve as a guide for development agencies involved in small-scale enterprise development. According to Parry, a good agent must:

- be well regarded in the business community and by the public sector; known to be honest, reliable and thorough;
- have good financial standing, with sufficient funds to put some equipment and spares on the shelf and to stand the cash drain involved in the start of a business, and during periods when sales are slow;
- have premises which are fairly central, accessible and easily found by customers, with facilities on the same site or closely adjacent, to demonstrate equipment and give basic training in its use;
- have a flair for business promotion;
- be familiar with the construction industry and procedures by which building materials are chosen for projects, and have a good perception of what ingredients are necessary for customers to do well in *their* business;
- have the technical ability to understand the operation of Parry equipment and the application of the products in order to give advice and service to customers.[16]

European and American firms with Parry's interest in genuinely appropriate technology can probably be counted on the fingers of one hand. Even if its profit margin was 50 per cent, the company would not be making a fortune for anybody. It lives on sales, but it thrives on innovation. It manages because it watches the bottom line and produces saleable products, but it grows because its employees genuinely enjoy what they are doing. They have produced a viable technology and the machinery to make it, but they continue to innovate, develop and improve the machinery, which is an important part of the business of staying in business.

On leading the blind

The projects described in this chapter are only samples of the building materials and construction technologies that have been developed in response to needs in the South. Much work has been done, for example, on introducing adobe construction to countries where it is not known, and improving it where it is. A wide variety of equipment to produce stabilized soil blocks – a compressed mixture of soil, sand and cement – is available, and many improved construction techniques for inexpensive housing have been devised using untrained labour and local materials.

Because low-cost building materials are intended to find their way into use among the poor, new techniques and technologies must be especially well tested. As with other products, some of the best successes have come about through the involvement of craftsmen and entrepreneurs for whom the technology is a next-step from what is already known. Parry's best customers are those already in the roofing business, and people familiar with similar materials. For some, a new material will be an addition to a product line, while others, such as Shelter Works in Nairobi, may discover that alone, one product is not enough to sustain a new enterprise. Thorough economic analyses of costs are necessary, as the fine pricing-line between the products of mini- and large-scale cement production demonstrate. But alone, a clear profit potential is not enough. Without good management, financial discipline and probity, even sure-fire investments – like brickmaking in Botswana – will fail.

The most successful projects in this chapter and throughout the book have usually flowed from some form of collaboration between researchers, government, NGOs and the private sector. In some cases, as with the Balaka Lime Project, the collaboration was planned from the outset. In others it was sequential. The Indian military research on mini-cement plants was taken over by formal research institutions, and dissemination fell largely to the private sector.

Government policies had an explicit role in the success of mini-cement plant development in China, but they have an important implicit role in the development or failure of other technologies. A roofing technology like FCR has shown itself to be cheaper than many of the alternatives. In Ghana, for example, it is 44 per cent cheaper than aluminium sheets. In Kenya, its installed cost is 40 per cent cheaper than concrete tiles. But because its installed cost is the same as that of galvanized iron, there might not seem to be much in the way of economic choice between the two. The production of galvanized sheets, however, is high in capital costs, creates one fifth as many jobs as cement tiles and requires three times the foreign exchange for every square metre produced.[17]

As with mini-cement plants in India, this suggests the need for a more pro-active policy stance by national governments in favour of the intermediate technology, if job-creation and import substitution are serious governmental objectives. Such policies might include:

- the purchase by government of FCR tiles instead of galvanized sheets wherever possible (the same applies to aid agencies – if NGOs simply bought their tiles from existing micro-entrepreneurs instead of setting up unsustainable new producers, the older ones would quickly become more successful producers and employers);
- the changing of building codes in appropriate cases to allow and encourage the use of FCR tiles where they are not at present permitted;
- the banning of inappropriate roofing materials, for example, 32 gauge (19 mm) corrugated galvanized roofing sheets are so thin that in tropical

climates they rust out within three or four years, at significant cost to users and – in the form of foreign exchange – to the country.

Other policies might include reductions in sales tax, the provision of business promotion services, and preferential interest rates for tile producers over those producing other less appropriate roofing materials.

The roofing tile case demonstrates two other lessons. The first is that entrepreneurs are not created overnight, and that great caution must be exercised in mixing micro-enterprise creation and new technologies. More important, people without experience in business should avoid the temptation to advise others without experience. In heavy traffic, the sighted generally make better guides than the blind. To use another metaphor, small vehicles can carry only so much freight. A project that is overloaded with too many objectives is much more likely to fail than one with limited objectives. Cheap, durable roofing tiles produced locally are one thing. No matter who owns the production equipment, jobs will be created. To expect, however, that a new technology with an uphill battle into a competitive market can also be owned and successfully operated by poor people without business experience and skills, may be asking too much of them, and of the technology.

CHAPTER XI
Light engineering and the very late starters

Mighty things from small beginnings grow

John Dryden

MASS PRODUCTION AND industrialization as we know them today were neither inevitable, nor were they the only way to structure industrial development. Modern alternatives or variations include the Japanese approach, using a wide range of small, flexible suppliers, and the industrial clustering found in Northern Italy, Mexico, Sialkot and Ludhiana – described in Chapter VI. This chapter will address another aspect of industrialization that receives too little attention in the literature on appropriate technology, and in thinking about manufacturing and industrial development – the role of light engineering.

Writers have characterized Germany, Italy and Russia as 'late' industrial developers, and Albert Hirschman has referred to Latin American countries as 'late late' developers, a title that could also be used to encompass what are now more generally referred to as 'newly industrialized countries' (NICs). There may be a further industrial category, the 'very late starters', those which by the 1990s had achieved little in the way of self-sustaining industrialization, and which, because of the advancement of NICs, start (or start over again) from a position of even greater relative disadvantage than was the case in the early years of independence.[1]

Most 'very late starters' are in sub-Saharan Africa, and unlike Asia and Latin America, most have an extremely weak light engineering sector. Before it became clear that they were *very* late starters, many of these countries accepted the trickle-down approach to growth, and unquestioningly equated industrialization with development. Eager to transfer the reins of their economies from European to African hands, and recognizing the limited ability of local entrepreneurs to finance industrialization at the pace or levels desired, many governments became directly involved in the process themselves, aiming to promote import substitution, create jobs, diversify the export base and diversify production away from the urban centres where it had been located in colonial times.

Given the thinking in the 1950s and 1960s about growth, development and the transferability of technology, there was no particular reason to think the approach would not work. Economic euphoria in the post-independence era was almost as heady as the political euphoria. Riding the wave of a world boom that began during the Second World War, many African economies were expanding at an unprecedented rate. In 1956, British Africa as a whole imported four times what it had in 1920 (at constant prices), while British African exports were ten times what they had been in 1937.[2]

Many countries embarked on extremely ambitious industrialization schemes. For political and ideological reasons, some, like Tanzania and Ghana, emphasized state ownership, while others, such as Ivory Coast and Kenya, welcomed foreign investors. Still others, like Nigeria, took a mixed approach. Although predominantly capitalist in outlook, countries like Kenya and Nigeria sought to keep certain industries in African hands, or aimed for levels of investment that were beyond what was available internally – much as Japan had a century earlier.

Report from the Institute for Advanced Studies in Hindsight

With hindsight, it is not difficult to see what went wrong and why. Many investments were unwise from the outset, but many did make sense in the prevailing economic climate – before the 1973 oil shock, and before the long downward price slide for tropical primary commodities. Part of the problem from the beginning, however, was that many investment decisions were made for political rather than economic reasons. Sometimes corruption was involved. Decisions were made without a full understanding of the economics of production, while still others were made on the basis of incorrect predictions about market growth or export potential.

A classic case of this was the establishment in Ghana of a mango-canning factory with an operating capacity that exceeded the entire world trade in canned mango products at the time. Bad decision-making in another case resulted in a Ghanaian trade delegation returning from a mission to purchase bottle-making equipment, having signed a contract for a plant that could manufacture bottles *and* plate glass. The minister responsible wanted the best glass-making plant available and thought plate glass might make a good export. The plant never produced plate glass and in later years was often unable to produce bottles.[3] The Zambian Government purchased the world's second nitrogenous fertilizer plant with a special delicate electrolytic reduction process based on coal; the factory quickly became a perennial and heavy loss-maker. A tyre manufacturing company, established on the clear understanding that it would operate with Zambian latex, was still importing rubber fifteen years later.[4]

One of the worst cases of bad planning was the establishment of a steel industry in Nigeria. The plan, initiated in 1975, called for a blast furnace and two direct reduction plants, based on domestic iron and coal. The first reduction plant was commissioned in 1981, and by 1985 there were 16 rolling mills in operation. A confidential government report in 1984 stated, however, that the project was uneconomic and would suffer large recurrent losses to the end of the century. By 1989 the project had cost $3 billion, and a further billion dollars were required for the completion of the first phase. Meanwhile the first reduction plant operated at one fifth of capacity because of power cuts and shortages of the *imported* iron ore that had become essential, due to underdevelopment of the local resource base.[5]

The state-owned Zimbabwe Iron and Steel Corporation, the largest manufacturing concern in the country, operated at a loss which contributed Z$493 million or 16 per cent to the country's budget deficit between 1980 and 1987, and ran at a loss of similar proportions throughout the 1990s.[6] In both Zimbabwe and Nigeria, the high cost of locally produced steel was passed on into the economy through other manufacturers who used their product.

These spectacular cases have their counterparts in hundreds of smaller investments. The backward and forward linkages of many industries were weak, and large plants stood in grand isolation from the rest of the economy. A sparkling new petro-chemical complex could turn out sophisticated products from a high-technology plant, and never purchase a single nut or bolt from the local economy.

Ironically, much of the manufacturing plant installed in Africa during the 1960s and 1970s – established to reduce imports – became highly dependent on imports for its functioning. These imports included spare parts, raw materials, foreign technical assistance and management. (Aid programmes are no different. Technical assistance accounted for an average of 29.3 per cent of all bilateral spending between 1981 and 1983. By 1997, the total had grown to 39.8 per cent.[7]) The domestic content of Nigerian cement production fell from 80 per cent in the mid 1960s to 40 per cent in 1975.[8] A heavy proportion of manufacturing was in consumer goods, and the import component was often especially high. Nigeria was self-sufficient in beer production in 1968, but by 1977, 41 per cent of production costs were accounted for by imports. Throughout Africa, few manufactured goods were exported, and as the prices of African primary exports declined, the cost of local manufactures increased in relative terms. For countries such as Ghana and Zambia, heavily dependent on a single export, the effect was catastrophic. Some plants closed, while in many others capacity utilization plummeted. Inefficiencies piled on one another; preventive maintenance declined, products deteriorated and prices continued to increase.

The result was disastrous in both real and relative terms. Investments had been squandered, jobs were not created, few imports were actually substituted in any significant measure, and the cost of manufactures usually did not fall, it rose. In 1986, manufactured exports from all of sub-Saharan Africa totalled $3.2 billion, less than what was exported in the same year by Malaysia alone. By 1996, statistical reporting on African exports of manufactured goods was so bad that the World Bank had figures for only ten African countries, and for four of these, the calculation was for a different year.[9]

One cause of the disaster was a poor fit between small domestic markets and the widespread enchantment with mass production. A further contributing factor was (and remains) a lack of choice in technology. Some industries operate at a fraction of capacity, not because of mismanagement or market miscalculations, but because the smallest plants available are far larger than the situation warrants. Added to this is the problem of supplier control of patents, and limited sources of supply. In the particle board industry, for example, machine suppliers will sell only complete, integrated packages. Some choices, however, were simply bad: the large and capital-intensive often won out over known, smaller, more appropriate technologies. Aid agencies were sometimes to blame, providing inappropriate advice or aid tied to inappropriate equipment. In some cases, corruption played a role.

But there are other historical reasons for inappropriate choices in manufacturing technology. One is that pre-independence industry in Africa, owned and operated by expatriates, focused to a large extent on industries with an immediate return rather than on the development of a long-term manufacturing capacity. This phenomenon has become more pronounced as high-tech firms set up factories in Asia and Latin America, and close them down just as quickly when tax breaks and cheap labour dry up, or when more propitious conditions present themselves elsewhere. Commercial capital was similarly dominated by foreign banks, which followed the pre-set industrial mix. Add these limitations to a more or less monopolistic pre-independence manufacturing structure and the normal development of a market economy is subverted from the outset. In Zimbabwe, for example, it is estimated that in 1982, half of all manufactures were produced under monopoly conditions, and a further 30 per cent under oligopoly conditions. Over half of all manufacturing output was produced by firms employing more than 500 workers.[10]

The tin men

There were two other cards in a deck stacked against African manufacturing industries. The first, the education system, will be discussed in Chapter XIV. The second was the virtual absence throughout much of the continent of a light engineering capacity. There was, and in many countries remains, a very limited scope for repair and maintenance beyond the modern factory gate. There is little capacity to manufacture spare parts locally, no machine tool industry, and virtually no engineering innovation beyond the most rudimentary. Each large-scale plant must either import its own repair facility, stock an excessive amount of spares, or hope for the best. By the late 1980s, the rule in many countries was to hope for the best, with the result that plants, or sections of plants closed until a single spare part could be flown out from the manufacturer.

For the development of appropriate technologies, this presents a special problem: in countries with a constricted capacity to produce and maintain even simple machines, and no serious innovative engineering capacity, the production and dissemination of many of the basic technologies described in this book are badly constrained.

Zimbabwean industry is far from typical of the African condition, partly because of its historically large white settler population and partly because of its three decades of strictly enforced import substitution and protection. Its relatively large formal manufacturing sector, although starved of new equipment and some raw materials after the mid-1960s, still looks more like a somewhat threadbare European economy than an African economy. With the number of jobs in the formal sector lagging tens, if not hundreds, of thousands behind the demand, however, the informal sector has grown dramatically. One has only to visit a place called Magaba on the outskirts of Harare to glimpse a part of Zimbabwe that is very much like other African countries. Magaba is a crowded, fenced-off area into which a large number of Harare's informal-sector producers were forcibly moved in 1989. The area has hundreds of stalls and shops, most without electricity. There are sign makers, painters, welders, carpenters, vehicle repair shops and used parts dealers. A few shops produce wrought-iron stoves, and some make chain-link fence with a simple device developed some years ago at the Rural Industries Innovation Centre in Botswana. There is a section where tinsmiths produce buckets, pans and implements from tin sheets. But beyond the odd bench saw and wood-turning lathe, there are few machine tools.

In recent years, a number of schemes have been developed by government and donors to promote the kind of small-scale enterprise that germinates in Magaba. USAID and NORAD have worked with the Confederation of Zimbabwe Industries, creating links between small and medium enterprises. The Small Enterprise Development Corporation, which once defined 'small' as having fewer than 50 employees and fixed assets of less than US$170 000, has relaxed its lending conditions and will provide loans of up to Z$10 000 without collateral. In 1995, ITDG established two service centres aimed directly at small-scale light engineering enterprises, enabling them to improve the quality and range of their capital goods production. As a result, more maize machines are being produced, more brick moulders and tile machines, all of which create new livelihoods for small producers.

Thanks to a particular donor interest, a study of Malawi's informal metalworkers, conducted on behalf of USAID in 1989, resulted in a better understanding there of the importance of metalwork, and led to clearer opportunities for action. The study[11] concentrated on enterprises that had a direct investment of less than $500 and gross annual sales under $4000. It found that, 'in terms of product value, the aggregate output of the informal metalwork enterprises probably leads all other producing informal subsectors except tailoring.' Forward linkages to agriculture, food-processing, transportation and consumption were particularly strong. There was also an important but undervalued backward linkage to Malawi's metal imports which eventually become scrap. Because the informal sector recycles scrap, each product is, in effect, an import substitute.

Tinsmiths produce watering cans, buckets, oil lamps, toys, stoves, pots and pans. Blacksmiths produce hoes, axes, trowels, crowbars and other tools, while metal fabricators produce hand trucks, window and door frames, wheelbarrows and ox carts. The most common metalworkers are tinsmiths, who often work in clusters in a market area. Although each tinsmith is independent, working in the same area with others has several advantages. Tinsmiths and their apprentices can learn new designs and techniques from others, for example soldering, or the use of a hand tool for knurling edges in order to provide strength and design qualities to the finished product. Buyers patronize areas where there is a good selection of products, and in a competitive setting, new producers learn quickly about costing, pricing, quality control and customer preference.

Nevertheless, the products of Malawian metalworkers are simple and the output is crude. In the Lilongwe Town Council Market, where some of the products are made and sold, metalworkers produce their goods in the open, seated on the ground. The Biwi Triangle, an area of shops built for small entrepreneurs by the International Labour Organization, has a more organized appearance, and is equipped with water and electricity. Here, among three or four dozen vehicle-repair shops, metal fabricators and carpenters, a man began experimenting with the casting of aluminium. In 1990, Frank Soko's tiny foundry was almost unrecognizable as such; his raw material was undifferentiated scrap, and his products were crude. But his was the first small commercial aluminium foundry in Lilongwe.

Frank Soko's skills and output were limited, but within months of start-up he was able to obtain an order from a Lilongwe brewery for a spare part that would otherwise have been imported. The brewery wanted 50 of the parts, but Soko was so encouraged by the order that he produced 150. Frank Soko was a pioneer and might well have contributed one of the small technological breakthroughs that are so important to wider industrial development and progress. While he overcame the technical and commercial challenges, however, there was one misfortune that overwhelmed him, closing his shop and extinguishing his efforts – AIDS.[12] The HIV scourge, which infects one in four Malawian adults, kills people, their knowledge, their potential and their hopes. It kills development.

'Doc' Powell and the Technology Consultancy Centre[13]

So far, except by way of quotation, only a few individuals have found their way into this story. This is more a problem of space than of design, for the real story of technology in history is the story of individuals – 'dirty fingernail people' like Frank Soko. Some of them are *champions* – zealots, idea thieves, pragmatists, innovators, inventors. The best-known have names like Fulton, Hargreaves, Edison, and Faraday, but behind them is an army of individuals like D.P. Saboo, relentlessly producing cement among his marigolds; or John Parry, making better and better vibrators in Cradley Heath; or Veljibhai Desai producing his Tinytech oil expellers in Rajkot. One of the most unlikely of these champions is an ascetic British engineer with a background in air-lubricated bearings, who went to Africa on a contract in 1970. Twenty years later, still in Ghana, John Powell was surprised to be awarded the Order of the British Empire. His Ghanaian friends were not. In fact many fully concurred with the mistaken upgrading of the award by the *Daily Graphic*, which reported that Doctor Powell had been knighted.

Powell had gone to the University of Science and Technology to teach, but to his surprise, in 1971 was appointed Director of the University's newly established Technology Consultancy Centre. TCC grew out of recommendations in an ITDG report, and in its early years it resembled many of the other appropriate technology institutions in the South. With assistance from ITDG and other NGOs, it experimented with a wide variety of appropriate technologies. After much trial and error – with technologies, techniques and implementing partners – some of its efforts were successful: glue, soap, textile weaving, honey.

But because Powell was a mechanical engineer, he was drawn to steel and iron, and to the machines that cut and shape metal. The Kumasi of 1971 was already much further advanced in its informal metalwork sector than today's Harare or Lilongwe. Erroneously called 'The Garden City', Kumasi – located 250 kilometres north of Accra – is a traditional crossroads for north-south and east-west trade, and lies at the heart of Ghana's cocoa-growing country. It was a logical staging point for traffic, and a thriving vehicle repair business grew up on the site of an old military magazine. When the town council moved the repair shops away from the centre of Kumasi, they took the name – Suame Magazine – with them. By 1971, Suame Magazine was a chaotic, sprawling place with 1600 separate owner-operated businesses, including 1100 workshops. Five hundred were involved in vehicle repair or modification and 300 were speciality shops offering welding, carpentry or blacksmithing.

Suame Magazine was a logical place for TCC to start, and an early survey homed in on a shortage of coach bolts used in the production of wooden truck bodies. Descending into a long nightmare of economic disaster, Ghana was already suffering from a perennial shortage of virtually everything imported. As a result, a few blacksmiths were making coach bolts to special order. But they were expensive and not very good. Powell had seen half a dozen centre lathes in the Magazine, used mainly for machining brake drums, and he assumed that if a process could be developed, higher quality nuts and bolts could be produced.

As these things go, it did not take very long. Within 18 months of opening, TCC had developed a process using mild-steel rod which was produced from recycled scrap at the Tema Steelworks. The rod was converted into a hexagon bar on a horizontal milling machine, then turned on a lathe to produce a shaft of the required length. Then it was cut with a small piece of the hexagon remaining to form the head. The thread was produced in a secondary lathe operation, and nuts were made by simply drilling and threading the same hexagon bar. TCC had the technology, and the nuts and bolts it produced in its own shop sold without any difficulty. It was to take eight full years, however, for the technology to 'transfer'.

There were three basic reasons. The first was machines. Although there were some lathe operators in Suame, there were no milling machines. This was an obvious drawback that would – in a carefully planned situation – have halted the project before it started. When TCC overcame that hurdle through an arrangement with ITDG to import used British machines and sell them to potential entrepreneurs, it discovered that most of the existing lathe operators were lathe operators, not entrepreneurs. Machining brake drums was the limit of their willingness to take technical and economic risks. Finding genuine entrepreneurs, therefore, became the second hurdle. And the third was the process. The first genuine entrepreneur abandoned nuts and bolts very quickly when he found that the process, developed in the campus workshop where he was trained, did not work in his shop. His workers' productivity (and therefore his profits) were considerably lower than had been anticipated.

The Adjorlolo brothers

It is worth following the story of these particular entrepreneurs and the milieu in which they operated, not only because of their eventual success, but because of what they represent for the development of a genuine African industrial base. Solomon Adjorlolo was a technician in Kumasi University's Physics Department when it began to produce standard resistance-plug boxes for secondary school laboratories. Seeing an opportunity in other teaching aids, Adjorlolo decided to set up his own business and in 1972 established a small workshop in Anloga, a part of Kumasi where other tiny carpentry industries were located. The two-person partnership was grandly named 'School Science Import Substitution Enterprises', or SIS for short.

With two wood-turning lathes, a small electric drill and a universal woodworking machine, Solomon and his brother Ben were producing 36 different items one year later, among them T-squares, tripods, ammeters, voltmeters, rulers and drawing tables. But as the bottom began to fall out of the Ghanaian economy, the market for cheap, locally made substitutes for imported educational items dried up. Desperate, the Adjorlolos took orders from TCC for benches and tables used in other projects. They manufactured a few broadlooms until cotton yarn disappeared, taking the weaving business with it. Finally, in 1974, in an effort to stave off collapse, they purchased a used metalworking production lathe, hoping to make metal jigs and cutters for other woodworking shops in Anloga. That too proved only a partial solution, and knowing that TCC was looking for an off-campus nut and bolt producer, they took a capstan lathe on loan and early in 1976 went into the nut and bolt business.

SIS's productivity was lower than TCC's, and the Adjorlolos gradually realized that they could make a better profit using the machines for other things. Having done some repairs for neighbouring saw-bench operators, they now saw an opportunity to produce spindles and bearing housings, and with a newly acquired welding set, they began to produce complete saw-benches. Not long afterwards, they produced their first wood-turning lathe, and within months, a dozen were in operation in Anloga.

The circular saw-bench and the wood-turning lathe were to the woodworkers in Anloga what the Spinning Jenny was to the eighteenth-century British textile industry. Within five years the area had been transformed from a place where carpenters and sawyers produced very basic furniture, into an area half a mile long where lathes whirled and saws screamed from dawn until well after dark. Products that had once been exclusive to large-scale formal sector producers were now being made by micro-entrepreneurs. And offcuts from the big sawmills – formerly a waste product – could now be put to productive use.

In the late 1970s, SIS began to turn its attention to corn milling, attempting to replicate imported machines which had become unaffordable. Import shortages proved the mother of invention, and eventually, with assistance from TCC, a corn-milling machine was developed which was in some ways better than the import. Used ballbearings replaced the brass bushings of the original, and broken half-shafts from trucks, made of better steel than the imported model, prolonged the life of the replica. In 1978, the SIS corn mill sold for half the price of the imported model. By 1985, with the Ghanaian currency all but worthless, the price of the local mill had increased eight-fold, but was still half the price of a smuggled model. And in 1991, under the full weight of structural adjustment, a fully convertible currency and totally liberalized imports, SIS was still competitive. A British corn milling machine cost $2100 in Ghana; an Indian machine cost $515, and the SIS model cost $400. In the 1980s and 1990s, SIS diversified into corn hullers, palm-kernel crackers and presses, cassava graters, potters wheels, block presses, and a combination circular saw-bench and planing machine. With 30 workers, SIS Engineering had crossed the line from informal to formal; from micro-, to small- to medium-scale enterprise. The Adjorlolo brothers had become an appropriate technology centre in their own right.

Although SIS was a spectacular example, perhaps because the two brothers had the right combination of technical and business skills, TCC has had many other successes with small engineering workshops, some of which became significant producers of nuts and bolts. These individuals, along with Solomon and Ben Adjorlolo, were genuine entrepreneurs; people who stayed with their endeavour, who adapted, who enjoyed what they were doing and who were able to roll with the punches and with the body blows that the turbulent Ghanaian economy threw at them. Some of their products were developed with the help of TCC. But in later years the tables were reversed, when Solomon became an occasional consultant to TCC, assisting it in the development of machines and techniques for other clients. He also acted as a consultant to government on small enterprise development. And his brother, Ben, produced a 40-page handbook on how to start and manage a small-scale business.[14] Something of a philosopher, Ben urged patience, saying that those who run fastest fall most heavily. 'A journey of a thousand miles,' he wrote, 'begins with the first step.'

John Powell identifies four stages of technology. The first and most primitive stage is characterized by hand tools that are not designed according to scientific principles. The fourth involves the use and development of scientifically designed automatic machinery. Much of Africa, including the tinsmiths of Malawi and the informal metalworkers of Zimbabwe, remains in the first stage, while governments and multinational investors have concentrated on leapfrogging into stage four. But it is only a parody of stage four; automatic machinery may be *used* in the formal sector, but it is often not used well, and it is certainly not developed indigenously.

Between the first and fourth steps, the second is actually the preliminary stage of development in mechanical terms; it involves human-powered machines designed in accordance with scientific principles. The next stage, Powell's third, relies on scientifically designed machines, powered mechanically. By the late 1980s, with SIS Engineering actually producing such equipment, and with hundreds of machine tools at work in Suame Magazine, Ghana had clearly moved from a position somewhere between stages one and two, to stage three.

Pumping iron

The importance of these developments to the genuine transfer of technology and the ability of a people to develop their own technologies cannot be overestimated. But in the mid-1980s, there remained a gaping hole in Ghana's metalwork story. Bronze casting had been known for centuries in West Africa and, unlike in Malawi, small-scale aluminium casting was reasonably well developed. But there were no iron foundries. As a result, metal fabricators like the Adjorlolo brothers often used scrap steel where iron would have been adequate. But as scrap steel became more expensive, the need for an alternative grew. In addition, some things, such as grinding plates for corn milling machines, had to be made from iron, and still had to be imported – legally or otherwise.

The irony was that scrap iron was plentiful, littering roadsides, workshops and farms where expensive imported tractors and heavy equipment lay rusting. Engine blocks were being used in Suame Magazine to fill giant potholes in the muddy tracks between shops. Almost a backyard technology in Asia, in Ghana iron casting was still *terra incognita*. TCC took up the challenge. Resurrecting a small crucible imported several years before, fuelling it with the palm-kernel charcoal used by local blacksmiths, and using a blower made from a knapsack sprayer engine, TCC melted its first charge of scrap iron in May 1985.

By then, TCC had learned that technology development and transfer were not going to take place on a university campus and had established what it called an 'intermediate technology transfer unit' (ITTU) in the centre of Suame Magazine, which by 1985 had expanded to some 40 000 workshops. There clients could make regular contact with TCC engineers for consultations on simple problems, for design assistance or for business advice. There, potential clients of a new technology, such as nut and bolt manufacturers, or potential buyers of a machine tool, could be trained in a production setting, sometimes for a year or more before going out on their own. And there, metalworkers saw for the first time what was possible with a ferrous metal foundry. By the late 1990s, when Suame accommodated 80 000 artisans, there were a dozen small foundries operating in the Kumasi area, the largest producing four times the volume of the country's largest formal-sector foundry.[15]

GRATIS

That TCC survived the 1980s is something of a miracle, given the chaos and unpredictability of Ghana's economy. But these were not the only problems. The University of Science and Technology had from the outset maintained an ambivalent attitude towards the idea of appropriate technology. Many academics felt that a university should be on the cutting edge of technology development, not wasting its time in a dirty place like Suame Magazine, developing second-rate technology for second-rate industry. Donor support was important to TCC's development, but it was often completely unpredictable. Early support from ITDG, Scottish War on Want, Oxfam, Oxfam Quebec and the EEC had been essential to several key projects. CIDA provided important technical and financial support, but there were long periods of famine while Canadian evaluators and consultants tried to figure out what to do between phases. USAID's brief fling with appropriate technology in Ghana resulted in a building and equipment for a second ITTU in northern Ghana, although an impenetrable wall of red tape caused the commissioning of the building to be delayed for six full years after the project agreement was signed. The ITTU did not actually become operational for another three years, making the nine-year total something of a record in project protraction.[16]

But it was government support that was most important, both for sustaining grants and – through some of Ghana's darkest days – for the morale and protection it provided. In 1987 this support manifested itself in a very important way. The government approved a new idea proposed by John Powell: the establishment of intermediate technology transfer units in several Ghanaian cities and towns, linked together by a central co-ordinating unit. The project, under the Ministry of Industries, Science and Technology, would be called GRATIS – the Ghana Regional Appropriate Technology Industrial Service.

With funding from the EEC, CIDA, GTZ and British DFID (then

ODA), the project began in 1987. The ITTU in Tamale, originally funded by USAID, was handed over to GRATIS, and the government contributed a large workshop at Tema originally constructed with Indian Government assistance. By the end of 1990, three more regional ITTUs had opened, and by 1996 there was a network of nine ITTUs throughout the country. Through consultancies and the sale of products from their own shops, the network covered a remarkable 56 per cent of their costs in 1998.[17] Although their efforts extended beyond engineering into agriculture, bee keeping, spinning, weaving, dyeing and food production, metalwork continued to be a major part of their operations.

Lessons from the metal trade

History shows that in industrialized countries, machine design, fabrication and adaptation have played a crucial role in the industrialization process. It was the development of the turret lathe around 1840, the universal milling machine in 1861 and the automatic lathe in 1870, that turned haphazard mechanical operations into a science, fuelling a second industrial revolution in Britain. Moreover, close links between small and large engineering firms have been critical to the processes of innovation and development. Without a local industry capable of testing and adapting technology to local needs, a country remains totally beholden to outside interests and influences.

Science is an important element of technological development. Many of the early inventions of the British Industrial Revolution worked *despite* a lack of scientific knowledge. Mechanically talented innovators succeeded as much through luck as through knowledge. In the second half of the eighteenth century, however, this changed, and most serious technical advances were essentially scientific.[18] Many of the successes of the informal sector in the South are similarly lucky, based perhaps on good mechanical knowledge, but without scientific foundation. In 1990, GRATIS began to deal with this problem by engaging a mining and mineral engineer to study small-scale aluminium foundries. As a student at the University in Kumasi, the engineer had done her final-year project in Suame Magazine, and had found, predictably, that aluminium casters there had no scientific knowledge of their trade. As a result, any aluminium scrap was used; pouring temperatures were usually too high; gating and venting – the way the aluminium is poured and how gases escape during the pouring process – were crude. Rejects were consequently high, and quality was poor.

With assistance from Valco, the giant aluminium smelter, GRATIS was able to obtain a pyrometer for measuring temperature, simple testing equipment, and samples of pure aluminium and silicon against which local products could be tested. Gradually some simple rules of thumb were developed for a training programme. For example, roofing sheets and drink cans, produced by extrusion or rolling, have a low silicon content and are therefore inappropriate for casting. More scientific approaches to gating and venting were developed, and simple techniques for testing the fineness, and the moisture and silicon content of the moulding sand were also developed.

As with the casting of iron, what may seem at first glance a very simple advancement was, in fact, a major step forward into a new type of industrial era. It bears out the importance of Schumacher's general view on the importance of *intermediate* technology – moving from the known to the next most appropriate level before making a mad dash for high technology.

The forward and backward linkages of Ghana's light engineering capacity are enormous. For agriculture, small machine shops are producing ploughs, planters and seed drills as well as more traditional agricultural implements. Forward linkages include the wide range of textile, wood and food-processing equipment being produced by SIS Engineering and other clients of TCC and GRATIS. Light engineering has important urban as well as rural implications.

Linkages with the formal sector are only beginning to demonstrate their potential. As is the case with the Malawi tinsmiths, many Ghanaian producers use scrap metal, and therefore have at least that much import substitution effect. But when they can produce good machines that are cheaper than imports in a liberalized, unprotected market, the import substitution effect is much more important. More crucial than that, however, items such as nuts and bolts, produced by the tens of thousands during the 1980s, were essential to the building of fishing boats and trucks and, in the form of wheel bolts, to keeping trucks and cars on the road when no imports were available.

When imports did become available, most producers were well enough trained to move into speciality production for hard-to-get or very expensive items. The ITTUs and clients of TCC and GRATIS are actually producing parts for large, formal-sector enterprises. One produces sprockets for a motorcycle assembly plant. Others have made special replacement parts for equipment in sugar mills and gold mines. Although structural adjustment programmes killed some small-scale enterprises, they have been unexpectedly beneficial to others. With funds available for rehabilitation of existing plant, large-scale manufacturers discovered that a local production capacity exists for specialized spare parts. The main GRATIS ITTU at Tema, for example, produced ten heavy cast-iron sprocket wheels with one-inch teeth for the chocolate factory, thus contributing to exports. The oil refinery, harbour facilities and ships in port have all ordered spares. The Ghana Water and Sewerage Corporation had funds to rehabilitate water systems throughout the country, many of which had been inoperable for years for want of spares. A key problem was brass pump impellers, which differ from pump to pump, depending on the head, the flow of water and the speed of the pump. Because the Tema ITTU had developed a pattern-making capacity, it was able to respond to a wide range of orders, a first step to introducing more complex types of metal-casting to other ITTUs and clients.

It might be useful to ask where the small-scale production machine shops and foundries might have come from, had there not been a TCC and a GRATIS. It is highly unlikely that they would have developed by themselves in twice the time, regardless of any special interest by government or donors in small-scale enterprise. In fact, had they not developed *before* the liberalization programme, liberalization might well have killed the opportunity instead of supporting what had developed. Credit was an important element in getting TCC clients off the ground, but it was one of the least important. The long and careful development, by specialists, of appropriate techniques and technologies was an essential part of the mix. It took eight years before the nut and bolt technology was adequately developed for transfer. Establishing six foundries took five years of thinking, and five more years of experimenting.

There are two other elements in the story of Ghana's success in light engineering. The first was the development of informal industrial areas like Suame Magazine. Smaller versions exist in most Ghanaian towns, and in other West African countries. Magaba, Lilongwe's Biwi Triangle and Town Council Market are very small, very pale versions of the same thing. The growth that comes from such clustering is not confined to engineering. As noted in Chapter VI, clustering is the hallmark of the Italian tile industry, Mexican footwear production and a dozen silicon glens, parks and valleys. It was equally true of seventeenth-century Dutch painting and eighteenth-century Austrian music. The African industrial areas recall the experience of the British Midlands, the Connecticut Valley and the Third Italy, and suggest that such clusters are not only an important element of learning and marketing, they may form an important building block on which diversification and greater sophistication can, with care, be constructed. This was certainly true of Anloga, with the advent of locally made saw-benches and wood-turning lathes.

In Ghana, the change in a place like Suame – from chaotic and cut-throat competition (in machining brake drums) to the beginnings of flexible specialization – would not have happened in only twenty years without catalysts like TCC and GRATIS. The catalyst within the catalyst, however, is the second fundamental element in the Ghana story: the presence, throughout the country's worst years, of a *champion* of intermediate technology and micro-enterprise development. John Powell's first twenty years in Ghana were enormously productive despite what seems, in retrospect, a long series of uphill battles with government, academics, donors and a perverse economy. He certainly did not create TCC or GRATIS alone; many excellent, dedicated Ghanaian engineers, social scientists and entrepreneurs worked with him over the years and supported key efforts when they seemed most likely to fail. And despite the problems with different institutions, there were many within government, the university and donor agencies who recognized what was happening, and who provided invaluable support at the right time. Without them it would not have happened. Equally, much of it, and perhaps all, would not have happened without Powell.

A reason justifying more discussion of 'champions' in development literature is that by leaving them out, the ultimate reason for a project's success or failure is often ignored. The counterpoint to this, however, is that apparently successful projects can be the product of one individual's enthusiasm, passion, or obsession. This 'Lawrence of Arabia Syndrome' can show immense promise while the Great Unifier is there, but things often fall apart once he or she leaves. This has not happened at TCC. After Powell left in 1986, there was an unfortunate slump in donor funding, but the sky did not fall; the institution continued in the capable hands of several Ghanaians who had been there with Powell for years. In addition to continuing with older activities, TCC went on to develop new and innovative projects in other fields.

The ultimate test of sustainability for GRATIS is time, and the support which it will continue to require from government and donors. Regardless of the future role of TCC and GRATIS, however, the seeds sown in Suame Magazine and other Ghanaian towns are already bearing fruit. In bringing together good people with diverse interests and attitudes, sound engineering skills and money, John Powell only proved what was already known by those who understood Ghana and knew something of the history of technology: that Ghana could make it.

PART THREE

An enabling environment

Since we've been overtaken by events, at least let's try and organize them.
Jean Cocteau

CHAPTER XII
Sustainability: myths and reality

*You take my house when you do take the prop
That doth sustain my house; you take my life
When you do take the means whereby I live*
              Shakespeare, *The Merchant of Venice*, IV, i, 376

Now the bad news

THE STORY OF GRATIS in Chapter XI is positive and encouraging, but it almost didn't happen that way. By the mid-1990s, GRATIS had two prominent funders, the European Union, which provided 5.2 million ECU up to 1995, and CIDA, which contributed C$8.56 million between 1988 and 1999. The two donors had very different ideas about sustainability. The EU supported GRATIS on the understanding that 'sustainability of the Intermediate Technology Transfer Units (ITTUs) does not automatically imply full "financial self-sufficiency".'[1] Contributions would be required for some time into the future to cover on-going development costs. CIDA, on the other hand, went into a second phase of support in 1992 on the condition that the project had to become self-financing by the end of that funding period.

The result was increasing pressure on each of the ITTUs to maximize cost recovery. This meant two things. First, it meant that GRATIS support was skewed towards those clients most able to pay (and by definition, perhaps, less in need of its support). And second, it meant that the ITTUs began to take production contracts, going into direct competition with the smaller businesses it had been established to support. To satisfy the god of sustainability, the underlying purposes of the project were compromised from the outset. A mid-term review found that 'the expectation that the GRATIS/ITTU network can become financially self-sufficient by the end of the project was unrealistic.' It went on to tell CIDA that it needed to 'clarify its policy objectives with regard to the achievement of development impacts within the context of an appropriate level of financial self-sufficiency.'[2] This is a bureaucratic way of saying that sustainability is about more than financial self-reliance.

Sustainability: origins

The practical problems of agricultural sustainability were discussed in Chapter VII. Because the issue of sustainability is so important to long-term development, however, it is worth digging into this complex subject further, going back to its origins in development thinking. Within a decade of the 1962 publication of Rachel Carson's *Silent Spring*, the environmental protection movement, hitherto polite and usually unobtrusive, had transformed itself into an environmental revolution. Aided in awakening public awareness by a series of environmental disasters, a new breed of writers warned of pollution, famine, a 'population bomb', the 'tragedy of the commons', in which common resources are depleted because no-one takes responsibility for them. Membership in environmental organizations grew dramatically – in the United States, Sierra Club membership tripled in three years – and more radical organizations such as Greenpeace and Pollution Probe took the debate to new and higher political levels. In May 1971, former CIDA President Maurice Strong commissioned Barbara Ward to write a report that would provide philosophical underpinnings for the upcoming UN Stockholm Conference on the Human Environment. The report, later published as *Only One Earth*, criticized international institutions for lacking 'any sense of planetary community and commitment'.[3] The *Limits to Growth*, published in 1972, announced that the roots of the environmental crisis lay in the problems associated with exponential growth, and its authors in the Club of Rome predicted catastrophe by the end of the century. While the *Limits to Growth* and other doom-laden texts were heavily criticized at the time, the hitherto unquestioned concept of exponential growth never again had quite the same resonance. The UN Conference on the Human Environment, held in Stockholm in the summer of 1972, recognized that a 'no-growth' approach would not be a viable policy option, but for the first time it made a clear and indivisible connection between the environment and human development. And because the conference was heavily attended by representatives from developing countries, Western environmentalists were forced to look at environmental issues with a global perspective.

The first real articulation of 'sustainable development' can probably be found in the 1980 *World Conservation Strategy,* published by IUCN – the International Union for the Conservation of Nature. Until then, there was no broadly accepted platform reflecting the environmental movement's change from protection to conservation, and there had been 'no reference base for reconciling the classical requirements of nature protection and those of sustainable economic progress.'[4] The idea, however, was more or less lost upon the 1980 Brandt Commission, which spoke only vaguely about new ways of looking at growth, and saw conservation primarily in terms of energy – mostly oil. Three years later, when it issued a follow-up report, *Common Crisis*, the Brandt Commission had moved environment up the agenda, devoting one half of one page to the subject. Writing in the fourth year of serious global economic stagnation, the Commissioners were much more concerned about a return to growth – almost any kind of growth – than they were about sustainable development.

This changed with the 1987 'Brundtland Report' of the World Commission on Economy and Development (WCED), *Our Common Future*. While some development practitioners are still reluctant to adopt a precise definition of sustainability, many accept the importance of the definition articulated by the WCED: *Sustainable development is development that meets the needs of the present without compromising the ability of future generations to meet their own needs*. This definition contains two key concepts:

- the concept of *needs*; in particular, the essential needs of the world's poor, to which overriding priority should be given, and,
- the idea of *limitations* imposed by the state of technology and social organizations on the environment's ability to meet present and future needs.[5]

Critics of the WCED definition argued that it was too narrow, that it had not come to terms with the growth problem, that it was too close to what some saw as a contradiction in terms: 'sustainable growth'. Others held that it lacked understanding of the real dynamics between poverty and environmental degradation, and that in the end it was unclear about what was to be sustained. If the WCED ducked the question of growth, it can perhaps be forgiven. Growth has become such a powerful shelter for economists, particularly development economists, that to pick a fight with the concept is to risk the widespread condemnation that befell the Club of Rome and its *Limits to Growth*.

In 1971, when he was a newly appointed senior adviser to the World Bank's Economics Department, Mahbub ul Huq made what was then a groundbreaking speech. He said that development economists had gone astray in at least two directions. 'First, we conceived our task not as the eradication of the worst forms of poverty but as the pursuit of high levels of per capita income. We convinced ourselves that the latter is a necessary condition for the former but we did not in fact give much thought to the interconnection ... Besides the constant preoccupation with GNP growth, we also went wrong in assuming that income distribution policies could be divorced from growth policies and could be added later to obtain whatever distribution we desired.'[6]

Mahbub ul Huq may have known the limitations of growth, but he did not reject growth *per se*. More than 20 years after his Ottawa speech, in a new incarnation as chief architect and editor of UNDP's *Human Development*

Report, he coined the phrase, and the idea of 'sustainable *human* development'. Not inexorably tied to environmental definitions, and not anti-growth, it offered some respite from the increasingly sterile debate that had swirled around the subject for several years, culminating in the 1992 Rio Conference on Environment and Development.

> Sustainable human development is pro-people, pro-jobs and pro-nature. It gives the highest priority to poverty reduction, productive employment, social integration and environmental regeneration. It brings human numbers into balance with the coping capacities of societies and the carrying capacities of nature. It accelerates economic growth and translates it into improvements in human lives, without destroying the natural capital needed to protect the opportunities of future generations. It also recognizes that not much can be achieved without a dramatic improvement in the status of women and the opening of all economic opportunities to women. And sustainable human development empowers people – enabling them to design and participate in the processes and events that shape their lives.7

The programme officer's reality

The idea of sustainable *human* development helped make the concept more accessible to development practitioners concerned with the implementation of sustainability on the ground in their day-to-day work. But the definition still included almost everything except the kitchen sink – employment, social integration, the environment, women, empowerment, growth. For practitioners, the idea of sustainability had a more immediate edge: how could outside interventions promote *self-sustaining* development? How durable will the results of an intervention be after the termination of external assistance? Was there, in the cold light of a programme officer's morning, any connection between a self-sustaining project and the mountains of paper from Rio, New York and Washington dealing with the carrying capacity of nature?

Practitioners have actually struggled with this dilemma from the outset of development assistance. It is the key question in all *ex post facto* evaluations. The sustainability of benefits was the subject of a seminal 1973 study of American aid, *We Don't Know How*.[8] Today, if we do not always know how, we do actually know more about project sustainability. UNDP has recently extended the idea, making it in some ways more understandable and more appropriate. UNDP is beginning to differentiate between what it calls 'static sustainability' and 'dynamic sustainability'. *Static sustainability* refers to the continuous flow of the same benefits that were set in motion by the complete programme or project to the same target groups. *Dynamic sustainability* refers to the *adaptability* of programme or project results to a changing environment, either by the original target groups and/or others. SIS Engineering and the entrepreneurs described in Chapter XI are good examples of this. As the fortunes of Ghana's economy changed, they were able to adapt and survive.

There are at least five critical aspects to sustainability in programming:

- technical sustainability (development and maintenance of appropriate trained personnel to meet needs at the local level);
- social sustainability (development and maintenance of community support as well as capacity within the community to play an effective role);
- political sustainability (development and maintenance of the political will necessary to sustain a major policy direction);
- managerial sustainability (development and maintenance of the capacity to direct and plan effective services responding to development needs);
- financial sustainability (provision of adequate human and material resources).[9]

Financial sustainability

Some of these factors are not new. Projects must be technically and managerially sound. They must have political support as well as local ownership – the basics on any feasibility checklist, although often more honoured in the breach than the observance. These will be examined in greater detail later. It is the idea of financial sustainability – as in the GRATIS case – however, that has the greatest appeal for project planners. It usually looms larger than all the other aspects of sustainability combined. Indeed, it might be asked why a project would be undertaken in the first place, if local institutions or beneficiaries do not have the capacity or willingness to finance it when it has concluded.

There are perhaps three answers to this question. First, a major part of the development enterprise is experimental.[10] If aid agencies knew how to end poverty effectively and conclusively, they would, presumably, have done so during the so-called first development decade. Much development work has been, and remains, speculative, and there is no reason to expect that the government of a poor country or its citizens would willingly pick up the tab for ongoing costs in a project that did not work, or that did not have value for them. If one made the generous assumption that only two-thirds of all development interventions are successful, as the World Bank does about its own projects, then there is little reason to expect financial sustainability in any more than that number.[11]

Second, local 'buy-in' is often weak. What may have started as a collaborative partnership between donor and recipient, can get carried away in the planning requirements of the donor agency, tied spending, and the sometimes overweening influence of consulting firms, experts and foreign NGOs. Donors, sometimes overwhelmed by the brilliance of their own insight on a particular issue, think of themselves as *catalysts*, showing the less prescient how things can or even *must* be done. Once the recipient organization recognizes that the Donor Knows Best, so the thinking goes, it will unflinchingly divert money from wasteful expenditures to the new donor venture. Donor arrogance has condemned more than one project to financial unsustainability from the outset.

Third, there is a basic problem of time and resources. British economist Roger Riddell argues that it is unrealistic to expect all projects to come anywhere near financial sustainability in the short three-to-five year time-frame of most donor interventions.

> In particular, it cannot be expected that distinct and discrete aid projects dealing with necessities (water, health, sanitation, medicines, school books and so on) will be able to be financially sustainable without funds that the direct beneficiaries are simply unable to provide. Equally, it cannot be expected either that the poor countries with chronic debt obligations, low savings rates, inadequate infrastructure and low levels of literacy will at one and the same time have the resources necessary to finance basic services, which its poor citizens are unable themselves to afford ... It is thus not practical in the foreseeable future to expect either the direct project beneficiaries or government budgets to be able to provide the funds for the whole array of projects which have been funded by aid donors.[12]

One area of recent and impressive success in financial sustainability is micro-credit. Some organizations, especially Southern NGOs, have demonstrated that lending to the poor, especially poor women, can be a financially viable proposition. High recoveries and market rates of interest have made organizations as far apart as Bolivia's BancoSol, the Kenya Rural Enterprise Programme, BRAC and Grameen Bank much more financially sustainable than would have been imagined even ten years ago. Money can be loaned to poor people, successfully recovered with interest, and loaned out again. This accounts, in part, for the very high appeal of micro-credit: it can reach the poor in a financially sustainable way. At the 1997 'Microcredit Summit' in Washington, Grameen Bank founder Mohammed Yunus went as far as suggesting that credit was a 'basic human right', and calls went out from the summit and the felicitously named organization that sponsored it – *Results* – to increase spending on micro-credit to $21.6 billion by 2005.

Apart from the sheer magnitude of the idea (this amount would have represented about 40 per cent of all ODA from all sources in 1996) there is something often missing in the discussion about micro-credit. The much-acclaimed results lean too heavily on high repayment rates and the financial sustainability of revolving funds. They rarely delve into what loans are used for, or who benefits from the profits – if any. Few ask whether it is possible for a destitute women to get herself *sustainably* out of poverty with a twelve-month loan of $200. If it were, it would be something of a miracle, and would say considerably more about the borrower than the lender. Closer examination is warranted. A very high proportion of Grameen Bank loans, for example, are used by women for traditional village enterprises – rice husking, cattle fattening, poultry. These were and remain activities that have always been carried out in Bangladeshi villages – by somebody. If work has shifted to the poorest, this is not a bad thing. But no *new* livelihoods are created by this type of lending. And in countries where micro-credit has worked most successfully, mainly in South Asia, it is the urgent need for *new* rural livelihoods that poses the biggest challenge to sustainable rural development. In Bangladesh, BRAC has made huge investments in developing such opportunities – village fisheries, dairies, a vertically integrated sericulture (silk) enterprise and entirely new approaches to poultry development and marketing beyond the village. But the experiments from which these achievements emerge are time-consuming, and for each success, there are inevitably failures. These, and the infrastructure needed to develop them, along with the social mobilization that precedes and complements a successful lending programme, cannot always be financially self-sustaining.

Lessons

Chapter XIII will have more to say about micro-credit. Two over-arching lessons, however, emerge from the debate on sustainability. First, sustainability is different from success; while sustainable activities in a development context may be successful, not all 'successful' projects are sustainable. And second, there is a difference between a sustainable activity and a sustainable organization. Another way of saying the same thing: sustainability is a relative concept. Management guru Charles Handy says that 'self sufficiency is an idle dream. Even those who cultivate their own organic plots need trucks built by others to drive along roads maintained by others.'[13] Few multilateral, bilateral and non-governmental agencies are sustainable without large amounts of regular external funding. They are supported, however, because they are deemed worthy of being sustained. Similarly, Southern institutions may not be financially sustainable, but they may be worthy of being sustained because the work they do or the services they provide are sustainable, or may be in due course. Ultimately, the key factors in sustainability can be clustered around four themes: stakeholder involvement, time and resources, capacity, and the policy environment.

Stakeholder involvement: ownership and participation

A 1996 OECD publication observed that, 'As a basic principle, locally-owned country development strategies and targets should emerge from an open and collaborative dialogue by local authorities with civil society and with external partners, about their shared objectives and their respective contributions to the common enterprise. Each donor's programmes and activities should then operate within the framework of that locally-owned strategy in ways that respect and encourage strong local commitment, participation, capacity development and ownership.'[14]

This is much, *much* easier said than done. At the micro level, one of the most difficult challenges in working with the poor, for example, is determining and understanding their needs. Many programmes are based on the needs of the community *as perceived by outsiders*, and are implemented based on the resources of the donor rather than on the resources and needs of the community.[15]

It is not enough, therefore, that people participate – they must genuinely feel that they own and control the process for their own development. Local structures should be used to help communities develop the capacity to undertake responsibility for local initiatives. Sustainability is enhanced when there is a commitment to the use of local knowledge and resources. Outside agencies need a close working relationship with beneficiaries in defining and solving problems with the resources at hand. A sustainable project respects and draws in local expertise, knowledge, and technologies. Robert Chambers observes that 'indigenous technical knowledge (ITK) is now respected more, and valued not only for its validity and usefulness, but because it is part of the power of the poor. ITK is strong on knowledge of local diversity and complexity, precisely where outsiders' knowledge is weak. In rapid change, its advantage over outsiders' knowledge is even greater.'[16]

Participatory planning may be time-intensive and may increase costs, but valuing learning and respecting the investment of time and resources by participants allows for more sustainable and lasting change. Of critical importance is the need for the full and active participation of women in local, regional and national programmes, and the promotion of gender equity through policy, planning and implementation. Sustainability is further enhanced when activities are phased and when projects allow for regular internal monitoring, and for evaluation that incorporates the perspective of the participants and beneficiaries.

Resources and time

It is a given that financial sustainability will only occur if there is planning for, and commitment to, the provision of long-term recurrent costs on the part of the local counterpart or beneficiaries. It is worth distinguishing again, however, between a sustainable activity and a sustainable organization. BRAC has made several thousand village groups financially self-sustaining through their earnings in a successful micro-credit programme. These earnings and other cost-recovery programmes may never be adequate to cover the full running costs of the overall organization. Long-term front-end subsidies, however, are creating financially sustainable *village* institutions, a primary purpose of international development assistance. A key is that community cost-sharing or cost-recovery schemes are most successful when management responsibility resides within the community.

Standard project time-frames (three to five years) are often too short for sustainability to take root; sustainable activities are complex, often involving a diverse set of stakeholders (traditional village and local structures, local government, local interests, the local development or community organization as well as national political and bureaucratic structures) – all of which can affect long-term success and sustainability.

Institutional capacity

Local institutions and their staff must have the capacity and credibility to maintain and adapt project benefits. The technical sustainability of primary health-care systems is dependent upon regular, comprehensive and ongoing training of heath professionals and volunteers. In order to improve the implementation of primary health-care and immunization projects, building of staff capacity in planning, animation, report writing, supervision, data collection and analysis must be ongoing, responsive and comprehensive.

USAID observes that the effectiveness of a partner organization is critical to fostering sustainability. In developing a partnership with an implementing organization, project planners and managers must consider a number of factors: the leadership of the organization, its experience in the sector, its current funding sources, its ability to recover its programme costs, the number of years it has worked in the community, and whether the organization has a strong local constituency. In essence, programme planners need to look to partner organizations (and leaders) with the following qualities.

- Do they have an entrepreneurial spirit?
- Are they motivated by the desire to create significant grassroots change?
- Are they pragmatic and do they have problem-solving skills?
- Are they strategic planners, able to develop management structures appropriate to carrying out their ideas?

Is there a reasonable target for project sustainability? Certainly, if sustainability is diminished by experiments, lack of local ownership, ill-conceived time-frames and projects that simply don't work, it is unreasonable to expect anything like full sustainability. Not to be confused with its two-thirds 'success' rate, the World Bank claims a sustainability rate in the neighbourhood of 50 per cent, with low rates of sustained benefits on technical assistance, and higher rates on infrastructure projects. Most bilateral donors, where the average is more like 25 per cent, find these figures – or at least the definition on which they are based – questionable.

There may be a justification for weak sustainability within the time-frame of the average donor project intervention. Riddell observes that 'in practice, where people and governments are poor and basic needs are being addressed by aid projects, donors tend to ignore or turn a relatively blind eye to their sustainability requirements.' The problem is not the desirability of working towards long-term sustainability, 'but rather the practice of using the same criteria for sustainability for all projects and programmes regardless of the particular context, circumstances or wider environment in which they are located.'[17]

The policy environment

Benefits are unlikely to be sustained if the national and international environments are inhospitable. This can mean different things in different countries. For example, fully sustainable human development cannot be achieved if people are denied basic political, social and cultural rights. Human rights are not *additional* to good development, they are integral to sustainable human development.

Sustainable activities take place more successfully in an environment where there is an effective government policy framework and an active civil society that can foster national and local debate, contributing to policy formulation through both participation and dissent. The Canadian Public Health Association, for example, identifies several policy-related factors in sustaining an immunization programme. Policies of government and other donors aimed at health improvements in general, can make a significant contribution to the sustainability of primary health-care (PHC) systems and immunization. Strengthening the sustainability of immunization and PHC systems requires a multi-disciplinary, integrated approach. Immunization alone, however, cannot be sustained without a functioning primary health-care system. Similarly, in water projects the provision of water must be seen as part of a broad-based understanding of, and effort towards poverty alleviation. Sanitation and hygiene considerations, for example, as well as economic factors, will be important to the sustainability of water projects.

Just as the host-government policy environment is important to the creation of a sustainable project activity, the global policy environment and the policy environment within development agencies are equally important. It makes little sense, for example, to assist developing countries with one hand, while blocking their textile exports with the other. Many have decried American arms sales to developing countries, which in dollar terms exceed official US development assistance. The OECD says this is a problem of 'coherence', and calls for donor governments to ensure that trade, defence and other policies support, rather than subvert development assistance. Part of the coherence problem relates to the old but unresolved issue of growth. Former World Bank maverick economist, Herman Daly, has this to say about growth and sustainability:

> Sustainability has had a hard time breaking into economic theory because the economics of the past fifty years has been overwhelmingly devoted to economic growth. The term 'economic growth' has in practice meant growth in gross national product. All problems are to be solved, or at least ameliorated, by an ever-growing GNP. It is the only magnitude in all of economics that is expected to grow forever – never to reach an economic limit at which the marginal costs of further growth become greater than the marginal benefits.[18]

Daly decries 'money fetishism' and the paper economy, joining a growing clamour for national accounting systems that include natural-resource depletion and the informal economy. He reminds us of what any planner understands at the project level when he or she looks at the population projections for cities like Calcutta, Lagos or Saõ Paulo: there truly are biophysical limits to growth. Good projects cannot flourish in a bad policy environment. The cause of development will not be furthered in an economy with a worthless currency and triple-digit inflation. Daly makes the same point on a broader scale: 'Sustainable development', he argues, 'necessarily means a radical shift from a growth economy and all it entails to a steady-state economy, certainly in the North, and eventually in the South as well.'

This chapter has touched briefly on many issues pertaining to sustainability at both the macro and the micro levels. If it has a message, it is that sustainability does not happen by fiat. It is a multi-faceted concept, not easily adapted to formulae and blueprints. Deciding how much of which elements are applicable in a given circumstance may be a complex matter, complicated further by new donor pressures to demonstrate concrete results, and by the problem that all donors have of attribution. What part of an achievement is the result of the donor's work, and what part is the result of a good partnership – flavoured with nuances of ownership, capacity and policy environment? What is reasonable under the circumstances? And how does the local connect with the global? There is no uniform answer to these questions, but in asking them and seeking answers in specific cases, policy-makers and programme officers may achieve more lasting and *sustainable* results than in the past.

CHAPTER XIII
Perspectives on women and technology

Home is the girl's prison and the woman's workhouse.
George Bernard Shaw

AT FIRST GLANCE, HISTORICALLY, women appear not to have been intimately involved in either technology or science. The degree of truth in this observation, however, depends to a large extent on one's definition of technology and science, and on who is observing and writing the history. It is not difficult to identify durable historical artefacts developed by men – from the Pyramids to iron and steel. Because women have traditionally looked after children, cultivated crops and preserved the harvest, woven and sewed clothing or tanned leather, much of their culture-bearing technological legacy cannot be put on display in museums. This has contributed to the invisibility of the important traditional and evolutionary technologies that women apply to food security, animal husbandry, seed preservation, horticulture, household health, energy, water and sanitation, and environmental protection.

By excluding women from academic and later from scientific education, by disenfranchising them politically and constraining them from the legal ownership of property, most societies have, until, recent years, erected and maintained powerful barriers to women's entry into enterprise, industry and technological development. Those women whose accomplishments actually found their way into the market-place and into historical records are, therefore, very special individuals: Catharine Green, who allowed her employer, Eli Whitney, to take full credit for her part in the development of the cotton gin; Mary Somerville, the nineteenth-century mathematician; Ada Lovelace, to whom the computer industry owes a great debt for concepts of programming 150 years ago; Marie Curie, the Nobel Prize-winning chemist.

In recent years it has been increasingly recognized that women are more than Adam's rib, more than an adjunct to productive society, more than simply mothers, cooks and drawers of water. They, along with men and children, *are* society. In industrialized countries, legal and social changes in the status of women over the past century have had an effect in more recent times on the behaviour of development agencies. In the 1970s and more so in the 1980s, thinking about 'women' moved beyond projects which aimed to make them better mothers and cooks. Income-generation projects no longer focused exclusively on health, handicrafts and sewing machines. 'Gender' – relationships between men and women and the role of women in society – began to replace simplistic thinking about 'projects for women'. Greater attention is now paid to the traditional knowledge of women, and to the need for new approaches that reduce their workload, reduce the amount of time required to carry it out, and provide new opportunities that empower them socially and economically. But the simple recognition of problems and past shortcomings, recognition that women's employment need not be stereotyped, that a woman *can* be a blacksmith, a welder or a lathe operator, does not solve the problems or create many genuine opportunities in such fields. There are still enormous social, technological and educational constraints involved in making even the smallest step forward.

In many ways, therefore, the role of technology in development is more critical to women than it is to men. Because while small technological changes can improve the lives of women, they can also unduly exacerbate the weakness of their position in society, and damage what fragile advantages they may already have. Transportation provides an example of this, and of the 'invisibility' of women's struggle with technology. Women spend large amounts of their day on the move: gathering fuel, carrying water, going to the fields, going to the market, taking children to school or to the clinic. In many countries, tradition excludes them from bicycles, their economic position excludes them from hired transportation services, and aid schemes exclude them from serious analysis. Priyanthi Fernando writes about an integrated rural-transportation project in Tanzania that promoted the use of donkeys and wheelbarrows. But women could not take a wheelbarrow to the fields and handle a child at the same time. Panniers designed for the donkeys could not carry firewood. The project made donkeys available for sale, rather than hire, so women – with less disposable income than men – had little access to them, and gender inequalities were reinforced.[1]

This chapter is *not* about the disparities in status, labour, income or access to services that women suffer. Nor does it belabour the fundamental truth that although women are major contributors to any nation's production of goods and services, a high percentage of their work is not reflected in national accounts. The important role played by women in agriculture, food-production, family income, and child and family health has been well documented over the past three decades. Neither does the chapter attempt to provide a comprehensive overview of technologies and techniques that have proved successful in improving the economic status of poor women. These are covered elsewhere. What follows is a discussion of *strategies* (rather than projects) aimed at advancing women into more productive lives, and at ensuring that they are part of the development mainstream, not an adjunct to it.

Options for women

The irony in attempting to discern alternatives for women in the development process is that many potential options are closed off in the very early years of a woman's life. In many countries girls receive less care and attention than boys. Their food intake is lower and their health is poorer. Although great strides have been made in recent years in education for girls, especially in Latin America and South East Asia, huge problems remain. Half the women of Africa are illiterate, and two-thirds of those in South Asia cannot read or write. While the gap in poor countries between the enrolment of boys and girls in primary school is only ten per cent, almost half of all children drop out before grade five, the majority of them girls.

For those women who do obtain an education, gender stereotyping is no less prevalent in the South than it has been in the North. In Ghana, for example, by the end of secondary school, 12 per cent of girls have opted for physics, chemistry and biology courses, and only 5 per cent are enrolled in mathematics. At university the imbalance continues: female enrolment in science faculties ranges from a low of 1 to a high of only 22 per cent. Mid-level Ghanaian technical schools have a 1 per cent female enrolment.[2] This plays itself out in statistics on female professional and technical workers. By the late 1990s in most industrialized countries, women occupied between 40 and 60 per cent of such positions, while in countries with a low UNDP human development index, the percentage – where statistics were available – range from 8.1 per cent (Niger) to highs in the mid-thirties.[3] The upshot is that women, whose needs are often greater than those of men, face the development challenge with fewer skills, less education and little spare time. And they often do so pregnant and in poor health.

In discussing strategic approaches to women and technology, a methodology has been suggested by two writers. Judith Tendler reviewed a number of 'better performing' Ford Foundation grantees, with a view to identifying why some were successful and others were not.[4] Her approach, using 'measures of significance' and applied to women's livelihood issues by Katherine McKee,[5] provides a useful means of focusing on the narrower issue of women and technology.

Tendler posited four measures of significance, the first of which is government policies and market forces. The second has to do with genuine and sustainable improvements in income. Following from this is the extent to which such improvements contribute to more general growth of employment and an expanded economic base, and whether or not beneficiaries have in some way been 'empowered' by the change. 'Empowerment of the poor,' Tendler argued, 'is central to their ability to increase their incomes.'

McKee breaks Tendler's measures of significance into three broad strategic approaches. The first, programming strategies based on a geographic area, usually encompasses a range of activities such as credit, marketing assistance, basic education, skill development and health programmes. The second – sector-focused strategies – deal with a particular industry or occupational area. The third, function-based strategies, can also be called the 'missing piece' approach, because it focuses on a particular constraint which, when removed, is expected to allow beneficiaries to advance. The most common 'missing piece' strategy today relates to credit, although there are others such as health or specialized approaches to education – improved and expanded primary education, or basic management training. Lack of credit, however, especially for the poorest, is increasingly seen as the major stumbling block to genuine, sustained income-enhancement.

The function-based strategy: too soon eureka

Within the function-based credit strategy, there is a now a vast army of purists who would limit external involvement to the barest minimum. This 'minimalist' approach suggests that credit *alone* is the missing piece in the development puzzle, and that little else should or needs to be offered. Stripped-down credit offers very limited training, no extension services, and no investment or marketing advice. This is essentially how the Working Women's Forum (WWF) of Madras, the Grameen Bank in Bangladesh and the Self-Employed Women's Association (SEWA) of Ahmedabad began, and selective evidence demonstrates that the approach works, especially for women. Four of Tendler's six 'better performing' organizations focused mainly or exclusively on poor women, and started with minimalist credit programmes. Three of them – WWF, SEWA and Grameen Bank – growing to significant scales of operation. By 1998, SEWA had 209 000 members, up from 15 000 a decade earlier; by 1999, Grameen Bank had 2.4 million borrower-members, up from 800 000 in 1990.

A relatively small minimalist project may help to illustrate the approach. Throughout the 1980s in Kenya, CARE operated a comprehensive, area-based approach to income generation for women in one of the country's poorest districts.[6] The programme began with a goat-rearing project and gradually expanded into training and the provision of grants for the promotion of agro-business, horticulture, bee-keeping and other small enterprises. But an internal review in 1988 confirmed what CARE already knew: most of the efforts were both unprofitable and patently unsustainable. Searching for a solution, CARE turned to the stripped-down credit model, adding only training in basic book-keeping.

The approach was simple. Basically, a women's group would approach CARE for assistance. After careful scrutiny and basic training for five elected group members, CARE would provide the group with a two-year loan, not exceeding KSh20 000 (then $800). Interest was set at 12 per cent. The fund was then managed by the group, usually on a consensus basis, making loans to individual members at rates of interest that are established by the group itself, ranging between 5 and 20 per cent a month. Loans did not exceed KSh2000 ($80) and each borrower had to have three guarantors.

Although still new and relatively small, by 1990 the CARE programme found itself with the same sort of results as larger organizations like SEWA and Grameen Bank. Despite the high rates of interest charged by the groups to their members, on-time repayment rates were in the neighbourhood of 95 per cent. More important was the discovery that most borrowers were making healthy profits on their loans, and many groups were likely to replace or exceed the amount of the original CARE fund within the two-year framework. Apart from such obvious successes, the minimalist approach proved to be inexpensive, easily replicable, and required little in the way of technical assistance. Moreover, it was capable of reaching large numbers of poor women.

This is news that donors love, and the micro-credit revolution that swept donordom in the 1990s saw considerable effort in fine-tuning the cost of developing on-lending groups, and calculating what rates of interest the traffic could bear in the quest for financial sustainability. By the end of the decade, mobilization cost estimates ranged from $100 per group to $400,[7] and feasible interest rates varied from 12 to 35 per cent, and sometimes higher. High interest rates are acceptable, because – as in CARE's Kenya project – there is gold in them thar hills. One study of 215 micro-enterprises in India and Kenya found that the average annual net incremental return on investment was 847 per cent, with 44 of the enterprises showing returns of more than 1000 per cent.[8] A superficial glance, therefore, seems to demonstrate that the CARE programme and others bear out minimalist claims. They seem to support McKee's observation that, 'Contrary to the predictions of many, the availability of credit has enabled self-employed people and micro-enterprises to stabilize and modestly increase their incomes, move into newer and more profitable lines of work, and build up a savings cushion.'[9]

Unfortunately, a more rigorous examination of the CARE programme and other frequently cited success stories reveals that this is only partially true. Many income-generation projects, especially those aimed at women, do fail because they are too ambitious. An external preoccupation with groups rather than individuals may be problematic. (As one acerbic observer put it, 'The rich get richer and the poor get groups.'[10]) In addition to income generation, many projects aim to reduce other gender inequities, to improve the general quality of life, to 'conscientize', to introduce a new technology, and so on. Many are founded on unrealistic assessments of what groups can do, and are based on weak assessments of the market in which women will have to compete. The pendulum, however, swinging from the complex integrated programmes of the late 1970s, now reaches the other extreme in its arc, and in (cheap) non-interventionist strategies finds answers that may not be there.

As with many other micro-finance organizations, virtually all the CARE loans were for trading. The women, many of whom were already petty traders, simply took their loan, went down the road to the next town where they bought maize, sorghum, soap or clothes, returned home and sold at whatever mark-up the traffic would bear. This was particularly successful in a district where village market-economies were weak and where the availability of capital was stunted. This project and hundreds of others have proved that poor women can be good credit risks. They *are* increasing their incomes and building a savings cushion. But there is little or no value added in what they do, no increase in productivity, no reduced dependence on external sources of supply. And to assume – all the studies in the world notwithstanding – that there are many opportunities where poor women can quickly repay a high-interest loan and make profits of 847 per cent, is to live in dreamland.

Women borrowers assuredly make the best of their loans, but their endeavours are limited to the narrow confines of the world to which they are generally restricted. This is why so many never get beyond traditional handicrafts, food-processing and petty trade. It would be a very unusual individual who could create a job or even a meaningful increase in income for herself with a $50 loan. Without tried investment opportunities based on sound technical information and *knowledge* (there is a distinction), minimalist credit programmes for women will mostly result in minimal improvements. When applied to technology, the functional approach has similar limitations. As examples in previous chapters have demonstrated, the development and display of an apparently 'good' technology seldom results in widespread uptake. Credit is certainly a missing bridge on the road to development, and technology is another. But alone they provide only part of the solution.

The sectoral approach

Genuine *productive* female job-creation and meaningful income generation remains an extremely difficult proposition. The BRAC poultry project in Bangladesh is a case in point. Going well beyond the minimalist and functional approaches, and actively seeking productive investments for poor women, BRAC devised and tested its approach to poultry rearing, and then built in training, supply and extension services. Before going ahead, BRAC workers learned as much as they possibly could about poultry. Essentially it was BRAC, not the borrowers, that created the productive income opportunity – for 1.2 million poor women as of 1997.[11]

There are three preconditions for a successful sectoral approach. First, a targeted sector should have large numbers of poor women already in it. In the BRAC example, one reason for success was that chickens were not new to women. Second, there should be a strong market for the product. And third, there should be government recognition of the sector's importance in order to promote supportive policy measures.

From a technology perspective, the approach may be further subdivided. Marilyn Carr suggests four distinctions.[12] The first is increased productivity of known or existing work through improved technologies, techniques and supporting services. A second is increased access to previously unavailable opportunities – through co-operatives, credit or technological change. A third approach involves the development of existing skills to produce new products or modified traditional products. And finally, a fourth involves new skills for new products or modified traditional products. Each subdivision suggests different complexities in training, institutional development and technological assistance.

The BRAC poultry project cuts across several of Carr's categories. It provides skills that modify a 'traditional product'. It includes new technologies (better chicks, vaccine), substantial organization, new support services and access to credit. And there is a strong demand for a product (chickens and eggs) that women already know.

Marty Chen offers a similar example from India, where traditionally, women have responsibility for the care and feeding of milk animals, and for the processing of milk products.[13] In the 1970s and 1980s, aiming to integrate women into the national dairy programme, a number of NGOs and parastatals in Andhra Pradesh began to develop all-women dairy co-operatives, training female extension workers and technical staff, and providing credit. By 1988, there were 23 000 active members in 400 women's dairy co-operatives and a further 73 000 in mixed co-operatives. In addition to improved income levels, the programmes gave women recognition and leverage (i.e. *empowerment*) at a policy level. And although these numbers are small in relation to the estimated 74 million Indian women engaged in dairying, the Andhra Pradesh success shows other organizations and other states what the potential may be.

Organization: smooth as silk

Food and post-harvest production are obvious areas that meet Marty Chen's criteria – large numbers of women are already involved, and there is a strong demand for the products. Another is the area of textiles and clothing. There can be very few organizations dealing with women that have not supported a spinning, weaving or sewing project. On a small scale, such projects can be highly successful. But replication – 'scaling up' to 'measures of significance' – has proven much more difficult.

A classic case of the scaling-up problem can be found in Botswana, where dozens of NGOs and vocational schools have, since the 1970s, organized women into weaving, knitting and sewing projects. School uniforms were often the primary output and there have been some striking successes. Bothakga Handknits, for example, which started in 1974, grew to 30 employees within ten years.[14] Starting with only two workers in 1973, Mopipi (Pty) Ltd had grown by 1990 into a company with 65 employees and an annual turnover of $300 000.[15] The Lobatse-based Designer Stitches, founded in 1985, grew into an enterprise with 21 workers within only four years. Established by local women, these companies and many smaller ones were formed in part as the result of a minimalist government policy, aimed at encouraging local enterprise with a generous package of grants and concessional loans.

As with other positive examples, these briefly described success stories in Botswana conceal a number of problems. Raphael Kaplinsky, referring to one of the cases cited above, says that 'the provision of finance alone has proved disastrous, since the owner lacked the experience and technological capabilities to manage such a relatively large enterprise efficiently.'[16] Debt has grown, and bankruptcy was at one point staved off only by direct assistance from the Swedish International Development Agency. For many other producers, the situation was worse. According to Kaplinsky,

> the effective freeing of the capital constraint on small-scale industry unleashed a flood of enterprises, more than half of which were in knitting and sewing, mostly providing school uniforms. Yet in the absence of effective marketing surveys, overcapacity was rapidly reached ... Moreover, despite the relatively simple tasks involved in these small-scale sectors, the technological and managerial capability of most of these small entrepreneurs was pathetically inadequate for the task at hand. Their rate of failure has thus been significant.[17]

Sometimes production bottlenecks are the barrier to scaling up. Silk and sericulture is an ancient village occupation in several Asian countries. High in employment creation and low in capital cost, it has the potential in many countries to generate regular income for large numbers of village women. A bottleneck in Indian silk production was identified in the reeling process, in which silk is spun from the cocoon onto a reel, or *charaka*. (The Indian *charaka*, also spelled *charkha*, was the technology on which Eli Whitney based his cotton gin in 1793.) New, improved *charakas*, developed by the Central Silk Technological Research Institute in Bangalore, and the Central Sericulture Research and Training Institute in Mysore, were placed in dozens of homes for testing.

In 1988, an Indian NGO involved in sericulture sought to evaluate the new *charakas*, and with assistance from ITDG made some surprising discoveries.[18] In a laboratory setting, the new *charakas* had demonstrated superior characteristics over the centuries-old village technology. In the field, however, it was found that the new model produced silk thread at half the rate of the traditional *charaka*. Further, the new model could only be used with the best cocoons – of the sort that had been used in the lab tests. As these represented only 35 per cent of the average home's supply, the new *charaka* was either abandoned or modified by the user. It was also discovered that the 'centuries-old' village technology had, in fact, been continuously adapted by users. Some *charakas* had ball bearings, chain drives, metal shafts, and a few even had small electric motors.

In Bangladesh, BRAC began to look into sericulture in the late 1970s, but rather than focusing on a single aspect of the process, took a more holistic approach. Working closely with the government's Sericulture Board, BRAC employees learned everything they could about silk, silkworms and the mulberry tree on which the worm feeds. They experimented with the raising of silkworms and the growing of mulberry trees. They experimented with spinning, weaving and dyeing, and with different field approaches, providing poor village women with silkworms and guaranteeing the price of silk. Land for mulberry trees being limited, government permission was obtained to plant along the verge of rural roads controlled by the Forest Department. By 1989, with almost 9000 mulberry growers and silkworm rearers, the programme had proven itself on a pilot basis and was ready for wider diffusion.

By the end of 1997, over 25 million mulberry saplings had been planted, producing an income higher than the minimum wage for the women who protect and grow the trees. The work was light, part-time, and usually not far from home. In addition, more than 21 000 women were working as silkworm rearers.[19]

At a cost of roughly a million dollars, the initial expansion may have seemed like a lot of money, but it translated into only Tk620 or (then) *$16 for the creation of a single, self-sustaining job*.[20] Essentially, BRAC had taken an old, known technology, and applied modern work-flow techniques to it, learning everything it could about sericulture, and ironing out bottlenecks during the development phase. Although not strictly a traditional activity involving large numbers of women in Bangladesh, the raising of silkworms was neither capital-intensive nor time-consuming. Trees, on the other hand, were well known to women, for whom they have always been a source of income and fuelwood. Here was Schumacher with all the i's dotted and t's crossed: income creation for thousands of people based on arrangements and a technology that was in every sense 'small, simple, cheap and non-violent'.

Many of the hurdles faced by women in attempting to escape poverty and improve incomes are the same as they are for any self-employed individual. Access to raw materials is essential. Credit may be an important factor. Government policies on price, marketing boards or agricultural inputs such as seeds and fertilizer are important issues. In the Bangladesh case, the access provided by BRAC to credit, silkworms, mulberry seedlings and training, as well as access to a stable market, were essential to success. Education, attitudes, choice of technology and luck all play a role. But for women, there is an additional element that plays a critical role: time.

Reducing the time required for food preparation or fuel gathering may allow a woman to devote greater efforts to income-generation activities, or possibly just a few hours less each week working. The palm-oil press in Ghana (described in Chapter VIII) did neither until improvements to other parts of the production process were made. The most attractive feature of many fuel-efficient stove programmes for women – who do the cooking – may be the amount of time, rather than the money, that they save. Any new activity must, therefore, be time-conscious. Women whose time is already filled with work will not be able to take advantage of a new idea unless additional time requirements are minimal. The 'improved' Indian *charaka* may have produced a better silk, but it took twice as much time to do so. The low amounts of time involved in Bangladesh mulberry tree planting, on the other hand, made it especially attractive to village women.

Cultural taboos

Socio-cultural constraints can be an important factor in the success or failure of new approaches for women. For example among the Ashanti in Ghana, the weaving of brilliantly coloured *kente* cloth is normally done by men and boys on narrow looms. Attempts over the years to introduce weaving projects for women failed because of a strong cultural taboo. Outsiders thought the taboo had to do with weaving, but in fact it had to do with the traditional loom, and when it was discovered that the restriction did not apply to broad looms, the uptake on weaving projects for women improved significantly.

Although it may be unwise, particularly for outsiders, to push too hard against apparently firm cultural constraints, it is perhaps worth remembering that in Europe and North America there were, at various stages in the first half of the twentieth century, 'strong cultural taboos' against women driving automobiles, voting and smoking cigarettes in public. (When asked if she thought women should be allowed to smoke in public, Gertrude Stein asked, 'By whom?') Until recent years in Bangladesh there were 'strong cultural taboos' against virtually every aspect of women's participation in meaningful income-generation. Women could not ride bicycles or motorcycles, and were therefore excluded from the field management of development projects. Women could not work in factories except as sweepers. And with the exception of a tiny, urban élite, women could not work in public at anything.

Many of these cultural biases — apparently not so strong after all — have fallen away in a very short period without significant opposition. It has been discovered, mostly through large-scale rural NGO projects, that women, although nervous at first, are quite able to ride bicycles and motorcycles to villages and (surprise, surprise), actually prefer it to the alternative. Men may grumble, but the world does not end. Bangladeshi NGOs and private-sector firms have also discovered that women can work alongside men in textile factories, pharmaceutical plants and even in metal fabrication shops without the sky falling. The CIDA-funded CARE Rural Maintenance Project has, since 1985, employed as many as 60 000 destitute women at a time in the maintenance of tens of thousands of kilometres of rural farm-to-market roads. The project has demonstrated that poor women can and will do such work, that the quality of their labour is good, and that this form of road maintenance is significantly more cost-effective than the way it was previously handled. While these projects have been specifically aimed at women, they have been carried out within a *gender* framework that views change as something needed by both women and men.

A Sarvodaya project in Sri Lanka, funded by the International Development Research Centre, trained women to produce handpumps in a machine shop. This non-traditional female activity may seem to over-stretch a point about avoiding gender stereotypes, until it is remembered who in the family is responsible for fetching water. Too many pump-maintenance projects have been placed under the control of men, often far away from the pump site. When pumps fail, as they often do, water must be found from an alternative source — usually the same distant, unhygienic source that the pump was supposed to replace. Many pump repairs require neither genius nor men, and could easily be handled by the women for whom the pump is an important part of daily life. The Jeevapoorna Women Masons Society in Kerala has a similar logic. Being responsible for day-to-day family sanitation and health, women saw both validity and opportunity in a Dutch project that sought to train them in latrine construction and cement-block production. Initiated in 1990, the project had supported the construction of over 53 000 household latrines by 1998.

Beginnings

Although there are many things women 'should not do', there are very few constraints on men. Marilyn Carr observes that 'the only rule appears to be that when a new technology which brings upgraded skills and higher returns is introduced, the men take over.'[21] Worse, in some cases it is taken over by the government. 'Empowerment' is therefore an important part of increasing the ability of poor women to earn better incomes. The formation of groups, co-operatives and other institutions that can speak on behalf of women and be heard, in a village or nationally, is a key part of changing the poverty–aid–technology matrix. Empowerment also flows from the social and behavioural skills, the creativity and the motivation that accompany group formation.

The question of empowerment is a complex one. Katharine McKee asks, 'is it more empowering to achieve modest, tangible gains through an incremental strategy, or to pursue a potentially higher reward, risking failure? And second, do the potential sociocultural gains justify the economic risks, especially for very poor women?'[22] One might extend this and ask if it is empowering when there are sociocultural gains but economic failure. The evaluations of many micro- and small-enterprise development projects for women often justify the investment and its subsequent economic failure on these grounds. A ginger pickling project in Thailand failed, but a case study finds consolation in the fact that women did earn some money while it functioned, that 56 per cent spent less time on household chores and a third spent one-third less time gathering fuel and water.[23] This sounds like the operation was a success, but the patient died.

It is worse when new strategies offer no improvement of any kind; when they simply disempower. Paddy husking in Bangladesh, once exclusive to women and one of the few ways a poor woman could earn cash income, has been taken over by mechanical and automatic mills in much of the country. In the 1980s, the new technology displaced women at the rate of more than 100 000 annually, and in the process, ownership, control and employment went to men. It is estimated that by 1987, 27 000 mills had displaced almost

35 million days of women's work annually.[24] The new technology was made economically attractive in part through the availability of subsidized electricity and cheap credit, most of it derived from donor agencies. Too often the choice of technology and the means of obtaining it is made by those least implicated in the outcome.

A broad-based study of women's economic projects in South Asia concluded that a broader perspective is needed on women's empowerment:

> This broader perspective should contextualize the concept in everyday lived reality: move beyond the (often) false dichotomies posed between economic versus political processes, and between individual versus collective power; offer concrete strategies for addressing everyday structural issues; and most importantly, move beyond the perspective of outsiders to incorporate the perspective of grassroots women.[25]

The study argues that the economic *is* political, and if successful, it will change power relationships in economic, social and political terms. Most of the successful income-generation projects that focus on individual borrowers and entrepreneurs also organize collective action for solidarity, mutual support, pooled savings, integrated supply and marketing services. In other words, no woman is an island.

A chapter on women and technology should perhaps end without firm conclusions, as so much is still being learned. Although there are areas of great promise, there are no simple prescriptions and there are none on the horizon. Successful interventions require sensitivity to the complexity of each situation. And if they are to succeed, they require professionalism of a high order – professionalism in research and organization, and professionalism in learning, listening and above all, *remembering*. And they require professionalism in applying what has been learned.

One of the key measures of a successful development strategy for women has to do with genuine and sustainable improvements in income. But the extent to which such improvements contribute to more general growth of employment for women and a wider economic base, is perhaps as important in the long run. Interventions that succeed in increasing a woman's earning capacity may well reduce poverty and even social inequality. But micro-level short-term increases in income and productivity can be wiped out by overnight policy changes. Institutional development, advocacy and policy reform – all ingredients of *empowerment* – are therefore legitimate and important objectives for any intervention aimed at improving the lives of women.

Those who are serious about the empowerment of women and who believe that the route to societal empowerment is through economic advancement have to be serious about means as well as ends. The introduction of new technologies without adequate backward linkages to supply and appropriate forward linkages into markets is as old a failing as it is objectionable, surviving only because of the stupidity of aid workers and the desperation of the poor. Hiding behind high repayment rates does not justify widespread donor inattention to issues of limited economic opportunity, weak sustainability and market failure. 'Sociocultural gains' conveniently listed under the heading of 'empowerment' do not absolve the designers of failed economic projects of responsibility for leading poor women – who are usually unambiguous about what *they* mean by economic empowerment – down a dismal garden path of 'other objectives'.

CHAPTER XIV
Employment and the informal sector: the economists lose control

> *That the developing countries cannot do without a modern sector ... is hardly open to doubt. What needs to be questioned is the implicit assumption that the modern sector can be expanded to absorb virtually the entire population, and that this can be done fairly quickly.*
>
> E.F. Schumacher

In 1973, Richard Jolly observed that 'Unemployment is increasingly emerging as the most striking symptom of inadequate development in most countries of the Third World.'[1] When he wrote that sentence, three-quarters of those who would be unemployed 25 years later were children or had not been born. By the end of the century, the situation was appreciably worse, and worsening. In the North, unemployment is largely a result of what economists call 'a deficiency of aggregate demand'. In the South the causes are often considerably more complex, involving not only inadequate demand, but inadequate supply, mismatches between opportunities, expectations and education, inadequate institutional mechanisms, and market distortions, both local and international.

Unemployment and underemployment have three prominent causes. The first, in rural areas, is land: land ownership and the employment it generates; security of tenure, the willingness to take risks and to adopt new technologies; size of land holding, which may create or reduce labour requirements. The second cause is a deficiency in human resources – in some countries both quantitative and qualitative. The third cause of unemployment – technology – is a mixed blessing. The introduction of modern technology has increased agricultural and industrial production, but in most cases it has done the reverse for employment. The negative employment impact of large-scale, capital-intensive technology, designed for industrialized economies and industrial workforces, has been a powerful contributor to underemployment in the South. Added to this problem is the widespread technology-related shift to synthetics in industrialized countries. This has dramatically cut into the demand for many basic exports from the South such as jute, sisal, rubber and cotton, lowering both foreign exchange earnings and the returns to agriculture.

The absolute poor in the South devote over three-quarters of their income – which is derived mostly from labour – to food. The low-paying, part-time daily-wage, and seasonal work to which they have access, are negatively influenced by a variety of factors. The poorest are more prone than the average to illness and disability, which in turn affects their ability to work. Child mortality among the poor is higher, resulting in more replacement births, which both increases dependency and removes mothers from the workforce.

Unemployment and underemployment are usually understated, especially for the poorest. The closer a person is to desperation, the more likely she or he is to work for whatever meagre wages or food are available. Also, because the poorest generally have the lowest levels of education, they tend to be the least employable. During seasonal fluctuations or periods of general economic hardship – bad seasons, bad years, drought, flood – these people become the most desperate.

Rural unemployment rates are generally lower than those in cities because of access to seasonal labour in the countryside, because of extended family networks and the availability of other productive assets such as a family homestead. Poor urban women are less likely to work than men, because more of them are married than their rural counterparts, and are less likely to head a household. Those who do work earn less than men, although not so much from wage-discrimination in the workplace, as from a gender bias in health, mobility and job-entry, and especially in education. This results in women being less well educated than men, and therefore less able to compete for urban jobs.

The employment situation is bad, and the prognosis for the future is bleak. During the 1990s it was estimated that 40 million new jobs would have been required in the South *every year*, simply to absorb those already born. In India, there were 10 million new entrants to the job market every year. In Indonesia there were 1.2 million and in Mexico there were a million, more than double the number of jobs created in those countries during the best years of the 1980s.[2] The International Labour Organization (ILO) estimates that at the end of 1998, one billion workers – one third of the world's labour force – were either unemployed or underemployed. The term 'underemployed' means 'working substantially less than full time, but wanting to work longer, or earning less than a living wage'.[3] Unemployment in the South is projected to rise, and it will be increasingly concentrated among the poor. The average annual growth of the labour force in low-income countries is 2.3 per cent, and in some countries it is closer to 4 per cent.[4] Yet labour demand has been increasing at the rate of only one-third to a half of one per cent each year. Urbanization and industrialization are reducing labour intensity, while both are raising the number of those seeking employment. The concentration of workers in large firms and in unionized public sectors reduces employment opportunities among new, poor, uncertified entrants. And inappropriate technological changes in the South and in the North only add to the problem.

The combination of poverty on the land and in rapidly expanding cities has led to the development of a vast unorganized 'informal' sector. The pathos of this 'teeming, motley circus' is summed up well by Paul Harrison:

> ... hawkers, tailors, whores, rickshawmen, market mammies, fortune tellers, watch-repairers, bush garages, one-man engineering workshops ... They require little capital, so anyone can set up in business. Their technology is labour intensive, often using improvised or second-hand equipment, creating the maximum employment. The raw materials they use are cheap and locally available, often recycled waste. Their products and services are low-cost, within the budget of the urban poor who for the most part cannot afford modern sector goods.
>
> Yet this great reservoir of savings, enterprise and inventivity has been, in the vast majority of countries, ignored or neglected ... It gets little or no official credit. Most of its capital is raised from family savings – the rest from moneylenders at exorbitant rates. It bumbles on, with inefficient management and ad hoc marketing. Indeed, it is often discriminated against or harassed. Official specification and standards rule its products out of many markets. City zoning and licensing arrangements make much of it illegal ... Informal enterprises often have to work in shantytowns, without light, power or water, sometimes even without a roof over their heads. Slum clearance programmes bulldoze away as many workplaces as homes.[5]

The 'informal economy' – neglected by planners and aid agencies, denigrated by economists and harassed by officialdom – represents as much as half the workforce in Southern cities, and a growing proportion of the rural workforce. The ILO estimates that in Africa, informal employment accounts for over 60 per cent of total urban employment. In Bolivia and Madagascar the figure is 57 per cent, in Tanzania it is 56 per cent, Colombia 53 per cent, Thailand 48 per cent and Venezuela 46 per cent.[6] The informal sector in Lima grew from 31 per cent of the economically active population in 1983 to 42 per cent in 1990,[7] and throughout the continent, 85 per cent of the new jobs created between 1990 and 1995 were in the informal sector.[8] According to Brazil's state statistics agency, between 1990 and 1997, the informal sector grew by 5.2 per cent while the formal sector contracted by 0.4 per cent. The reason, according to the head of the agency: 'bigger firms using more machinery and fewer workers ... the excess workers are absorbed into smaller endeavours.'[9] In 1998 South Africa's statistics department revised its estimate of national informal sector employment from 1.2 million to 1.8 million, a 50 per cent increase.[10] This is not the 'marginal' thing that economists often disparage, nor is it a passing phase. It is the economic lifeblood of millions of families and the countries in which they live. It is also the story of employment through 99 per cent of human history.

National development strategies in most developing countries concentrate on agricultural production at one end of the development spectrum, and urban-based industrialization at the other. Between these poles lie the small-scale informal-sector 'entrepreneurs'. These unsubsidized people have contributed significantly to keeping national economies afloat while subsidized industry and the formal sector stagnate, and in some cases collapse around them. Carpenters, metal fabricators, bakers, soapmakers, weavers, tinkers, tailors and candlestick makers, they are the backbone of the 'underground economy' or 'black market'.

Take Karachi, for example. There, 53 000 of the city's 79 000 new housing units each year are constructed by and for the informal sector. Sewers, waste management and other services are also managed by the informal sector. Solid-waste removal provides employment to more than 40 000 families. Karachi's public transit system meets a fraction of the need; the balance is provided by 13 500 minibuses, operating without bus terminals, depots or workshops, and mostly without official sanction. There are 42 000 small workshops and businesses in Karachi's slums. In one, Orangi, there are 400 private (i.e. informal) clinics against only 18 government health facilities. There are 700 private schools, ten times more than are provided by government. Even warehousing for the formal sector is handled informally. As much as half of Karachi's warehousing is done in the old city where buildings have been torn down and replaced by ground floor warehouses with men's dormitories above. Over 76 per cent of the city's population works in the informal sector.[11]

Operating beyond the control of administrators and out of sight of the statisticians, the informal sector exposes the irrelevance of much economic planning. It also underscores the need for greater recognition of the fundamental right to economic initiative. Maria Nowak makes the case succinctly and passionately:

> Despite considerable rhetoric about satisfying basic needs, fighting poverty and, more recently, growth with equity, in many Third World nations, misery is still on the rise, fuelled by high rates of population growth and structural adjustment policies, which do not always have a human face. This rise in misery, which also reflects a major failure of foreign aid, cannot be checked through assistance and charity. However it would certainly be attacked more effectively if, apart from an environment more conducive to self-help and microenterprise development, socialist and market-oriented governments alike would recognize the individual's right to economic initiative by providing access to capital, training and technology.[12]

Capital

Despite its size and its significance to the economies of most countries, the record of domestic and international assistance to the informal sector is poor. Many countries have established specialized organizations designed to offer a range of support services to small enterprise, but their output is generally low, and the focus is usually on credit, combined with a limited approach to training. Success is often measured in loan repayment rates rather than in productive job-creation. Between 1978 and 1983, one such body in Tanzania established only one surviving small industrial enterprise each year for every two staff members. Ghana's Entrepreneur Development Programme – under the National Board for Small-Scale Industries – was given a mandate to assist micro-enterprises, but activities consisted mainly of raising false expectations, with credit applications in some cases taking more than a year to process. Most of these organizations do not reach the poorest elements; they are often prohibited from working with unregistered, 'informal' enterprises; and few have the mandate or the capacity to deal with the broader macro-economic and legal climate that make such enterprises both possible and necessary. In Peru, only one-third of one per cent of all credit was channelled to the informal sector in the late 1980s.[13]

Subsidized interest rates, introduced to assist the poor, have often attracted the better-off who are more familiar with the value of money. Rather than assisting the poor, who are used to paying very high interest rates, subsidies have often done the opposite, putting low-interest programmes into the hands of the wealthy. A further problem is that small enterprise-oriented credit operations are often biased towards trading operations rather than producers, because petty traders are more likely to be profitable, and can start making repayments quickly. Productive enterprises must often therefore resort to the family for financial assistance.

Education

The relevance of education to the development of local technological capacity, to national economic growth, to improved equity and to poverty reduction has been incontrovertibly demonstrated. Detailed studies conclude that education contributes significantly to growth, and that returns on education are comparable with, or more than, returns on physical capital.[14] Education is significant in reducing poverty and improving income distribution, and the contribution of *primary* education to growth and income distribution is much more significant than for most other levels.

Four years of education can be directly correlated with better agricultural output – an average of 8.7 per cent in 31 countries surveyed, and over 20 per cent in Nepal.[15] Primary education means better health and nutrition, improved child survival and lower fertility. On average, a ten per cent increase in girls' primary school enrolment will decrease infant mortality by

4.1 deaths per thousand, and the same increase in secondary enrolment will reduce mortality by another 5.6 deaths per thousand.[16] For micro- and small-scale enterprises, primary education means an ability to follow written instructions, to measure accurately, and to understand some of the basics of science and organization. But these facts, known and widely reported for years, have not resulted in any appreciable shift in official development spending: in 1996, 10.8 per cent of all bilateral assistance was spent on education, but less than 2 per cent of it went to the primary level.[17]

Despite the dismal aid performance, great improvements in enrolment and literacy rates have been achieved over the past three decades. Adult literacy among African men rose from 51 per cent in 1980 to 66 per cent in 1995. For women the number rose from 30 per cent to 47 per cent. Of the 37 countries for which there are comparable data, 24 were spending more per capita on education in 1995 than they were in 1980 and 47 out of 65 were spending less on the military.[18] There are still huge problems, however. In some countries the quality of education, crowding, absenteeism and the availability of books and supplies is worsening. Perhaps the greatest loss in investment – for children, parents and society – can be found in high drop-out rates. In Pakistan, for example, only 48 per cent of the children who enrolled in primary school in 1990 were still there in 1995.[19] If the high drop-out rate for Pakistan and other countries was disaggregated by gender, it would undoubtedly show that the problem of already low rates of enrolment for girls is compounded by a disproportionately high drop-out rate.

For those who actually get beyond primary school, the level and type of employment associated with higher levels of education has changed dramatically since the 1960s and 1970s, especially in Africa. Because of the inability of the formal job market to keep pace with the number of school leavers, the education system has actually been turning out more and more graduates for *self*-employment, without explicitly recognizing the fact, and without adjusting the curriculum to better arm students for the inevitable. And where tertiary education is concerned, in many countries, the poor are effectively excluded. In Ghana for example, the richest 20 per cent of households benefit from 45 per cent of subsidies to tertiary education while the poorest fifth received less than 6 per cent. In Malawi, the richest 20 per cent take two-thirds of the subsidies while the poorest receive only one per cent.[20]

Over the years, vast sums have been spent on vocational training, although much of it is impractical, and for many students (and employers) it is regarded as a second-class means of obtaining a diploma, rather than as a key to employment. Many of those in the productive informal sector learn their skills through apprenticeship rather than in school. A detailed study of micro- and small engineering enterprises in Ghana's Suame Magazine, for example, found that 84 per cent of the proprietors had received their training through apprenticeships and that only 16 per cent had any formal technical education.[21] While apprenticeship has a long and honourable history as an educational tool (witness Peter the Great's experience as a carpenter in a Dutch shipyard), it has severe limitations in today's world. Analysis, experimentation, knowledge of basic science and measurement are more important to the development and adaptation of technology than ever before.

But in much of the South, technical education is regarded as it was in Victorian England — something for the lower orders. An Englishman returning from the Paris Exhibition of 1867 noted that 'France, Prussia, Austria, Belgium and Switzerland possess good systems of industrial education . . . and England possesses none.'[22] Somehow this Southern throwback to Victorian England has to change; a *premium* must be placed on technical education, as it was in European countries eager to catch up with and surpass Britain's Industrial Revolution.

There have been experimental efforts to do this. Mapfure College for ex-combatants in Zimbabwe gave technical and business training to young adult would-be entrepreneurs, and then worked with them in the establishment of co-operative business enterprises in carpentry, construction and metal fabrication. The Botswana Brigades, village-based youth polytechnics in Kenya and the 'education with production' experiments in Zimbabwe and Botswana have attempted to equip students with genuine skills for employment. The Malawi Entrepreneur Development Institute (MEDI) has added a strong entrepreneurial element to an institution that started basically as a vocational training institute.

In the past, many of these programmes or projects were small in scale, were tied to the enthusiasm of a single individual, or were crudely designed to keep people in the rural areas rather than to enhance the scope for enterprise, income or human fulfilment. The most successful projects, such as the Technology Consultancy Centre in Ghana and GRATIS, work with people already in the labour market, and provide a larger package of inputs, including access to credit, raw materials, machinery and follow-up business advice. Selecting individuals from those already in the market has the advantage of reducing technical training costs and increases the likelihood of assisting people who are genuine entrepreneurs. GRATIS and TCC in Ghana, and MEDI in Malawi, all have rigorous tests in order to ensure that their 'students' have the right mix of motivation, analytical skills and problem-solving ability.

Kenneth King argues that this approach, replicated and vastly expanded, must be the educational approach of the future:

> Explicit promotion of the enterprise culture may help at the margins, but . . . if enterprise and self-employment is only promoted through education and training

institutions, there will be little effect. A really major set of changes in the enabling environment and in the financial support of microenterprise will be necessary to alter the dominance of education and training for employment towards some significant measure of education and training for *self*-employment.[23]

Writing about the American educational system, which he believes is mired in outdated form and content, Robert Reich might be describing the most urgent educational requirements of the South. In order to increase the stock of problem-solvers, innovators and entrepreneurs, he argues that the formal education system must concentrate on the creation of four basic skills.[24] The first is a capacity for abstraction, an ability for individuals to interpret and rearrange for themselves the chaos of information that surrounds them. The second is a capacity for 'system thinking'. Reich describes system thinking as an ability to see reality as a system of causes and consequences, as discovery of opportunities that lie in the relationships between cause and effect. The third is a capacity for experimentation, a willingness and an ability to forego some of the stultifying security that is engendered by textbook-bound educational systems, and to act on the newfound capacities for abstraction and system thinking. And finally, if it is to equip individuals for innovation, problem-solving and leadership, an educational system must be able to foster in people an ability to collaborate, to work in teams, to share problems and information.

Technology

While minimalist credit can be an important start, it has serious limitations for the longer-term growth and development of both individuals and industry. But credit alone, whether minimalist or otherwise, is often as far as small enterprise development programmes go, partly because it is an obvious need, and can be easily 'evaluated' by examining loan-repayment rates. One reason for the limitations of such programmes is that tested investment opportunities for the micro-level borrower simply do not exist in many countries. Often it is incorrectly assumed that credit, having provided the wherewithal to obtain a welding set or a cassava grater, has solved the job-creation problem. But, as examples in this book have shown, working with new tools, new materials, and new production techniques creates new problems, and is not something that should be left to chance, especially where an impact on the poor is anticipated.

Although the need for integration between credit and credit-worthy opportunities is widely acknowledged, there is an unfortunate and widespread programming mismatch between credit and technology: most established credit institutions have little interest in technology, while many NGOs and technology institutions have no interest in the micro-entrepreneur. The

Technology Consultancy Centre and GRATIS in Ghana are exceptions, although finding donors who would contemplate a marriage between credit and technology was not a simple matter for them. TCC originally approached the problem by importing reconditioned British machine tools and selling them at cost to its clients. GRATIS started a hire-purchase scheme which gave its clients access to machinery that existing credit institutions would not finance. Similarly, in Zimbabwe, ITDG created two centres for the small-scale light-engineering sector where entrepreneurs could rent (and later purchase) modern equipment and receive training in new technologies and business development.

Policy

There is a blithe assumption in much development writing that good policies will result in growth, and that growth will result in more employment in the formal sector. For most bilateral and multilateral aid agencies, 'good policies' include macroeconomic stabilization, liberalization of trade and internal markets, and institutional reform. Stabilization, however, may not be in the best interests of the political establishment in a given country, and where the primary emphasis by external creditors is placed upon trade liberalization and reduced government involvement in the economy, anything can happen. The Congo/Zaire – under Mobutu and afterwards – has staggered along despite an apparently never-ending process of state decay. William Reno puts the endlessly collapsing state phenomenon under a microscope in Sierra Leone and finds an explanation: given the cost of the alternatives, state decay can be an effective means for a ruthless politician to retain power and acquire personal wealth. Between 1967 and 1985 the government of Siaka Stevens allowed the formal state apparatus to gradually unravel so that control over rivals – both political and economic – could be exercised through less formal but personally more effective means. The creation of a 'shadow state', based largely on diamonds, worked well. It worked better when the DeBeers mining operations were nationalized, and it worked better still a few years later when the IMF insisted that the diamond trade and other government businesses be denationalized. Several of 13 major state-owned enterprises went to politicians and to the Lebanese diamond dealers who – with Stevens's active participation – were purchasing the country's major resource on the black market. As Reno puts it:

> Both IMF and World Bank agendas presumed the existence of a border between state and private enterprise that would insulate the entrepreneurial energies of businessmen from politicians' interests in order to generate greater taxable economic activity. The World Bank staked $53.9 million on this presumption to 1987 for the rehabilitation of enterprises slated for privatization.[25]

Among other things, Stevens retained personal control over import licenses. He allocated them to his diamond-exporting cronies who then parcelled out scarce commodities to petty traders in the informal sector. By 1989, after two sets of 'reforms', 70 per cent of all salaried workers in the country were civil servants, while in the so-called 'private' sector, 'each big-time operator had spun a web of small traders in a way that defies any laws or methods of free capitalist enterprise.'[26] Further 'reforms' and further loans from the African Development Bank, the IMF and the World Bank did not create growth, did not restore the formal economy, did not prevent further decay. Nor did they prevent the eruption of a brutal civil war after 1992, which in the following seven years killed 75 000 people and created more refugees than Kosovo.

Collapsing and predatory states aside, few enterprise development organizations, especially those interested in the informal sector, take a strong interest in policy development. Some economists suggest that even in more solid economies, informal sector enterprises do not graduate into the formal sector, do not grow, and are not a source of economic dynamism. In some cases – perhaps a large percentage – this is true. Many individuals in the informal sector are not there from choice or because they are 'entrepreneurs'. They are there because they are desperate.

This does not diminish the importance of the sector, however, nor does it detract from the need to take it far more seriously. Nor does it deny the fact that many informal-sector industries, like SIS Engineering in Ghana, do 'graduate'. Many would probably enter the formal sector if they could – in order to gain greater legal and economic security, and to take advantage of official credit markets. But in his now famous Peruvian exposé, Hernando de Soto demonstrated the high cost of crossing the barrier. His researchers in Peru established a fictional, nine-worker garment factory and applied for registration: the exercise required 11 permits, consumed 289 days and cost the equivalent in cash and time of thirty-two times the monthly minimum living wage. Permission to occupy land and build a house on state property required 207 bureaucratic steps and the involvement of 48 different government offices. At a week for each step, the process would have consumed almost seven years.[27] The same exercise carried out in Tampa, Florida took 3.5 hours, and in New York, 4 hours.[28] Unable to afford the cost of entering the formal system, informals therefore remain outside, reluctant to invest in new technology or expensive equipment, unable to gain access to mainstream markets for their sales, borrowing from informal credit sources at exorbitant rates, often producing poor quality goods, while evading taxes, building codes, and labour regulations.

In the late 1980s, interest in the informal sector and micro-enterprise development began to take off, with micro-credit providing an alternative source of investment capital for the informal sector. Micro-credit reached new levels of public recognition with a large 'micro-finance summit' held in Washington in 1996. By the following year it was calculated that there were 7000 institutions reaching an estimated 8 million people with small loans, certainly a revolution of sorts. But Grameen Bank and BRAC together had over 4.5 million members in 1997, most of them borrowers. This means that the other 6998 institutions were reaching perhaps 5.5 million people. The implication is that while many of the other institutions had certainly become large, many were still very small, reaching a few hundred people at most. As Malcolm Harper, a pioneer in micro-enterprise development puts it, 'Although the total achievement is remarkable, only a minute proportion of the total market has thus far been reached ... The time is now ripe for established commercial banks, which have the necessary physical, financial and human resources, to enter the market.'[29]

The cost of informality

The costs to people, to governments and to society of an unsupported, unregulated, anonymous informal sector are enormous. Among those identified by Hernando de Soto are:

- declining national productivity: formal-sector industries spend undue amounts of time, energy and money complying with official regulations. In an effort to avoid costly labour regulations, they use inappropriate technologies which require more capital and fewer workers. The informals – undercapitalized, dependent on high rates of interest, devoid of legal facilitating instruments, and usually suffering from an excessive ratio of labour to capital – are also inefficient;
- reduced investment: the insecurity of the informal sector and heavy regulation of the formal sector weigh heavily against investments in both;
- an inefficient tax system which overburdens the formal sector;
- increased power and water costs for the formal sector (the use of half of Lima's water and electricity, for example, is unaccounted for);
- stunted technological progress: in the absence of property rights, protection for technological innovation and access to cheap credit, a major part of the productive sector remains technologically retarded;
- difficulties in formulating macroeconomic policy: the absence of accurate data on the scope and size of the underground economy introduces a considerable degree of uncertainty into the design of macroeconomic policy.

Although small-enterprise development has gained popularity among donor agencies in recent years, attention outside Latin America generally remains focused on small industry development corporations, often ineffective in their stated objectives, and less effective in reaching the poorest of the micro-entrepreneurs. The conventional approach still compartmentalizes support into credit, technical assistance, and vocational and business training. Important as these are, integrating, legalizing and promoting the informal sector, and providing it with access to the formal market-place, are more critical. For this to happen, an appropriate official climate for micro-enterprise development is necessary. This includes simplifying the legal and bureaucratic requirements for registration and reporting, and a democratization of access to services such as credit, water and electricity. And it requires decentralization. Decision-making authority for approvals, loans and other services must be in the hands of officials in direct contact with the problems that a policy or funding mechanism seeks to address. Better local policy analysis on the informal sector is also urgently needed. The formal business and banking communities must be drawn into the challenge, and NGOs could become more involved in policy analysis as well as in actual programme implementation.

Marilyn Carr notes that, 'while donors express a theoretical preference for the macro-policy approach, external assistance rarely takes the form of policy assistance, and support for small enterprise is rarely a consideration in policy leverage situations.'[30] The challenge, therefore, for governments, NGOs, donor agencies and the formal private sector – if there is to be encouragement of enterprise, meaningful development of local technological capabilities, and a reduction in unemployment – is enormous.

CHAPTER XV
Globalization, adjustment and all that

> *My critique of the global capitalist system falls under two main headings. One concerns the defects of the market mechanism ... The other concerns ... the failure of politics and the erosion of moral values on both the national and the international level.*
>
> George Soros

MOST APPROACHES TO appropriate and intermediate technology start at the bottom, with projects. This book has attempted to demonstrate the importance of *context* to the efficacy of such projects, and to show that a country's policy framework is critical to the success or failure of an intervention. If anything is clear from history and from the cases described in this book, it is that the development and application of appropriate technology must be approached from the top as well as from the bottom if it is to succeed.

A period of adjustment

The 1980s, sometimes called a 'lost decade' for the South, was a time of difficult adjustment to new economic realities, adjustment that overflowed like a tidal wave into the 1990s. More than adjustment, it was a wholesale retreat from the failed and often mismanaged governmental *dirigisme* that characterized the 1960s and 1970s. The retreat was a forced march, a new kind of *dirigisme*, stage-managed by multilateral and bilateral donors, innocently labelled 'policy dialogue'.

The adjustment was a logical accompaniment to a remarkable confluence of events that began in the mid-1970s and ran through the 1990s: the oil shocks; debt crises; a period of deregulation and privatization in Western Europe and North America; a political climate characterized by what has been called 'Ronald Thatcherism'; the collapse of communism; the 1995 creation of the supra-national World Trade Organization and the 1999 beginning of the Millennium Round of trade negotiations; the revolutionary change in information and communication technologies. Taken together, the process and the phenomenon have become known as 'globalization'.

In the South, formal 'structural adjustment' packages became one of the handmaidens of globalization. In the parlance of international financial institutions, structural adjustment means the comprehensive reform of macro- and micro-policies to rectify inappropriate policies that hamper economic performance. Adjustment is required as a response to various 'shocks' (oil crises, collapsing commodity prices) which work against standard development objectives such as employment, economic growth, poverty reduction and improvements in the balance of payments.

The term 'inappropriate policies' can be a euphemism for many things. In general, however, it refers in this context to excessive state intervention in the economy: investment in state enterprises, misguided price controls, and high levels of protection in support of inefficient import-substituting industry. It includes overvalued currencies which favour the import of capital goods over local manufacture, while reducing traditional exports.

The prescriptions, usually in the form of policy conditions set out in the adjustment lending programmes of the World Bank and the IMF, and widely supported by bilateral donors, include varying degrees of currency devaluation, reductions in government expenditure the better to reflect real resources, and the bringing of domestic prices of key products, such as petrol and fertilizer, into line with world prices. It involves trade and exchange liberalization as well as liberalization of domestic prices, and a variety of 'institutional reforms' such as the closure of inefficient government enterprises, or their sale to the private sector. By reducing the role of the state, it is believed, inefficiencies will be reduced. Likewise, 'rent-seeking activity' – an anodyne euphemism for corruption – will be curtailed.

It is believed that by reducing opportunities for rent-seeking – through trade, price and exchange liberalization – and by replacing state inefficiencies with private entrepreneurship and the market-place, the disastrous effects of decades of control and mismanagement will be swept away. Devaluation and trade liberalization will further improve efficiency and reduce poverty by 'tilting incentives back in favour of food and export agriculture, and by promoting labour-intensive industrialization.[1] The previous chapter described the naivety of this assumption where countries like Sierra Leone, Zaire/Congo, Nigeria and Liberia have been concerned. There, liberalization, privatization and reform actually improved opportunities for 'rent seeking'. But that is perhaps a quibble.

There are other problems with adjustment. When structural adjustment programmes were first introduced in the early 1980s, they were seen as short-term undertakings; short-term pain for long-term gain. Although adjustment would involve hardship, it was like chemotherapy. The cure would kill some healthy cells, but in the end the cancer would be gone.

Whether there has been more damage to good cells than bad in this process has been the subject of great debate for almost two decades. The pro argument says that without adjustment, all the development gains of previous years would have been endangered. Anti-adjustment rhetoric argues that it places all the gains of previous years in danger. The pro-adjustment argument states that protectionism and state intervention – which sought to develop local capacity, protect infant industry and increase employment – created instead, a false, uncompetitive economy that led to fewer and fewer state options and greater poverty for more and more people. The anti-adjustment position argues that attempting to alleviate poverty by cutting government expenditure and throwing economies open to the market does precisely the opposite, inducing massive lay-offs and forcing cutbacks in health and educational services.

No world order

In many ways, the debate about structural adjustment became almost irrelevant in the mid-1990s because its theologians, noticing that their religion had practical limits in the tide of globalization rising around them, rushed to practise themselves what they condemned in others. The British government's attempted 1999 £152 million bail-out of the BMW-owned Rover car company, for example, aimed essentially to protect 9000 jobs.[2] The British government knew, just as other industrialized governments do, that in an era of fast-moving capital and volatile markets, those jobs could easily have moved to Vietnam or Malaysia or the Czech Republic, following other jobs that had already left. Ironically, in today's world of less government, more government of the right sort can be very helpful to business. Take, for example, the State of Kentucky, which in 1993 gave General Electric $19 million in tax breaks over ten years in order to keep a washing machine factory open. Or the State of New Mexico and Sandoval County, which gave Intel Corp. benefits worth one third of a *billion* dollars to establish a new computer-chip factory. Or the State of Alabama which gave Mercedes-Benz incentives worth $253 million to establish a plant. Or Ohio, which in 1997 helped General Motors to save $30 million in taxes and provided a $1 million cash grant in order to facilitate a plant expansion – this to a company whose profits for the two previous years totaled $11.8 billion.

Some call this globalization – the threat and the possibility of taking the factory elsewhere. *Time* magazine called it 'corporate welfare'. 'The US government alone', it reported in 1998, 'shells out $125 billion a year in corporate welfare, this in the midst of one of the more robust periods of the nation's history.'[3]

If heavy industries like BMW and Mercedes-Benz are that fortunate in attracting government subsidies, and if they are that mobile, then lighter, more sophisticated industries are even more fleet of foot. In the new globalized economy – made possible by the revolution in communications technologies – textiles, chemicals, pharmaceuticals, the semiconductor industry and electronics firms all locate and relocate where the incentives are greatest, and where labour is quiescent or unimportant to the overall mix. Technologies and capital equipment are more likely to be imported than locally made; more likely to relate to the needs of the firm than the needs of the country's longer-term development. Rather than government placing conditions on the investor, it is the investor that places conditions on government.

Fast-growing competition for capital greases the skids of this kind of globalization, along with dramatically compressed time-frames. Manic investing has served to divorce finance from real economic value, and speculation in national currencies occurs at unheard of speeds and volume. Market mood swings are frequent and volatile, as the Mexican debt crises in 1982 and 1995, the Asian meltdown in 1997 to 1998, and half a dozen other panics in the second half of the 1990s demonstrate.

Ironically, in some ways there is less globalization today than there was 125 years ago. Labour and migration are less mobile now than they were in the late nineteenth century, and free trade reached much higher levels in the 1860s and 1870s than in the 1990s. An IMF report suggests that international capital movements as a proportion of economic output were, in 1997, well below what they had been in the 1880s.[4] The difference today is that capital movements are more speculative, more leveraged, more nervous, less patient. Charles Calomiris, professor of finance and economics at Columbia University, observes one of the results: in the 43 years between 1870 and 1913 there were five major banking crises with only one leading to an exchange rate crisis. In the last two decades of the twentieth century, however, there were 90 major banking crises, many of which led to runs on national currencies. 'Capital rushes around the globe much more quickly than ever before, and is more prone to sweep into countries and then out. Technology has created an electronic herd but as yet no electronic cowboys to control the herd.'[5]

Those who suggest that there should be 'cowboys to control the herd' – electronic or otherwise – are quickly dismissed as anti-business simpletons. George Soros, one of the most successful speculators of the 20th century, puts the shoe on the other foot, likening those who resist improved transparency and some form of global regulation to religious *fundamentalists*. Market fundamentalists, he calls them.

> According to market fundamentalism, all social activities and human interactions should be looked at as transnational, contract-based relationships and valued in

terms of a single common denominator, money. Activities should be regulated, as far as possible, by nothing more intrusive that the invisible hand of profit-maximizing competition. The incursions of market ideology into fields far outside business and economics are having destructive and demoralizing social effects. But market fundamentalism has become so powerful that any political forces that dare to resist it are branded as sentimental, illogical and naive.[6]

The purpose here is not to debate the details of globalization or structural adjustment, but to suggest that infatuation with the market and the notion of 'getting the prices right' to the exclusion of all else, including social and political objectives, equity, employment, poverty alleviation and environmental issues, is naive and simplistic. It is especially simplistic in its view that more jobs and an increase in manufacturing exports will naturally flow from 'right prices'.

Unfettered market forces may well promote employment and efficiency in general, over long periods of time in advanced market economies. But they don't work as well in climates where markets fail, where inappropriate industries – often monopolistic – are already in place, where capital is foreign, skittish, and only there for the short haul. What is so odd about the confident macroeconomic prescriptions made out to patients in the South is that macroeconomic theory is itself in a state of confusion, and neoclassical theory, apart from its distance from important political realities, has serious and obvious drawbacks as a basis for policy recommendations in the South. In addition, the pressures that donors apply to encourage economic policy change in the South may be incompatible with their efforts to foster democracy. Bilateral and multilateral financial institutions – largely free of public scrutiny and accountability themselves – have enormous influence over 'sovereign' governments and the lives of millions of people. Insisting on the absolutes of 'perfect competition', they have fallen heavily under the sway of market fundamentalists. 'Whenever politics and business interests intersect,' Soros says, 'there is a danger that political influence will be used for business purposes.'[7] These 'business purposes' have certainly gained the upper hand in the thinking of many donor agencies.

Market fundamentalism and the tenets of its faith notwithstanding, Southern economies suffer from a number of severe non-market 'distortions' that affect the supply and demand of goods, services, and especially technology in perverse and negative ways. Economists should not, therefore, be allowed to get away with the idea that where the advancement of technology is concerned, economics are all that matters. Technological change has also to do with attitudes, ideas, individuals and institutions. Getting the prices right is necessary, but alone, it is not enough to promote optimum technological change. Getting the priorities right is also essential.

Socioeconomic distortions

If the problems which adjustment programmes address – price fixing, subsidies, overvalued exchange rates and inflation – are set aside (i.e. cured), there remain a number of socioeconomic distortions that strongly affect the choice of technology. Different social groups, for example, have different access to technological change, to markets and to skill development, as other chapters in this book have demonstrated. Land tenure patterns can have a serious effect on the adoption of new technologies because of the need for economies of scale required to justify even simple scaling up from manual techniques to animal traction, or the introduction of a tubewell.

Distortions caused by attitudes are a particular problem, especially in Africa and Latin America. Large-scale production is often equated with modernity and development, while locally produced goods, the informal sector and small-scale labour-intensive technologies are regarded as inferior, or even as a plot by the North to continue its subjugation of the South. Two examples, again from Ghana, illustrate the problem. During the 1970s, the Technology Consultancy Centre developed locally made equipment for soap production. The only large-scale soap manufacturer, Lever Brothers – heavily dependent on imports for its production – had all but ceased operations by the mid-1980s, encouraging many small entrepreneurs to invest in the new technology. The trade liberalization that came with structural adjustment, however, allowed Lever Brothers to resume operations, and within months, the small soap manufacturers were out of business, their savings and livelihoods gone.

The cause was not price or quality; the local soap was cheaper and the quality was identical to that of the more popular Lux. The cause was attitudes, shaped by massive advertising campaigns, attractive packaging, perfume, and a pathological belief that foreign is better. The pervasiveness of this attitude can be seen in Ghanaian prices of the ultimate global consumer icon, Coca-Cola. If there is a single international product whose manufacturer prides itself on the uniformity of its quality, it is Coca-Cola. In the 1970s, the company left India in a much-publicized high-level squabble rather than compromise the quality of its product. Yet in Ghana, in 1999, locally produced Coca-Cola sold for 800 cedis a bottle, while imported tins sold briskly alongside the bottled variety, for 1500 cedis – an 87.5 per cent premium for 'foreign'.[8]

Another example of the inability of markets to deal with technological choice is demonstrated in the world's electricity supplies, divided between 110 volt and 220 volt systems. Part of the world drives on the right, part on the left. There are broad gauge and narrow gauge railways, often in the same country. Steam and electricity were abandoned in the development of automobiles before the technologies that might have made them profitable had been properly developed, but railways are still powered by diesel and electricity, and in some cases by steam.

Further attitudinal problems are posed by the phenomena several writers have referred to as 'engineering man', 'economic man' and 'bureaucratic man'.[9] Engineering consultants, whether they work for government, donor agencies or the private sector, generally favour the most modern, technically efficient equipment which, in the international market-place, is usually capital- rather than labour-intensive, is frequently urban-biased and of an inappropriate scale. Economists, unfamiliar with the technical aspects of a proposed solution to a problem, are understandably prone to accepting the 'best' engineering advice.

By the time a proposal reaches 'bureaucratic man' – an individual who is usually technologically conservative and traditionally averse to altering the status quo – the options have been substantially reduced and the decision is all but made. (Gender-specific references here reflect the fact that most technological decision-making is still heavily male-dominated.) What influence he retains will normally be biased in favour of reducing the workload. This places a premium on economies of scale rather than smaller projects which are more complex and require more time, more judgement, and greater risk. Risk avoidance, common to engineers, economists and bureaucrats alike, results from low tolerance in government for deviation, non-conformism and failure. But risk avoidance encourages inappropriate and, in the final analysis, inherently more risky investments. That engineers, economists and bureaucrats all veer naturally towards large-scale, inappropriate technologies is demonstrated in a comparative study of the very different development strategies of Tanzania and Kenya, which found much more similarity than difference in their technological choices.[10]

A final socioeconomic distortion is worth mentioning again: corruption. Corruption is not unique to any particular time or place, as those who follow Belgian, Japanese and American (or any other) politics and business know. But it has a particularly pervasive influence on the choice of technology in the South, especially where purchases from the North are concerned. Southern officials are often insecure and badly paid. Decision-making is slow, files are lost and forms are unavailable. Procedures are interminable and simple transactions become impossible. In Brazil, obtaining an export licence in one documented case required 1470 separate legal actions, and involved 13 government ministries and 50 agencies.[11] Hernando de Soto's example of registering a fictional small garment factory in Peru is similar. The pointlessness of the lengthy and costly exercise was demonstrated by the fact that none of the authorities involved ever realized they were dealing with a simulation.[12] Poverty, insecurity, rigid adherence to rules combined with a reluctance or an inability to delegate, a weak rule of law, and an obsession with status, all lead to obvious temptation and predictable results. A Westinghouse nuclear plant scandal in the Philippines, a Bofors uproar in India, a Nigerian cement scandal, French arms-for-timber swaps with Liberia's up-and-coming rebel leader Charles Taylor – all are examples of what can happen when unscrupulous salesmen and buyers come to town.

Institutional distortions

Institutional distortions introduced by foreign equipment sellers are one thing. Foreign investors introduce another kind of technological distortion. As interested in reducing risk as the individual bureaucrat, corporate investors (often in the form of multinational corporations) will, in the absence of pressure to do otherwise, import a known technical package, rather than attempt to develop something more appropriate locally. The known, usually made for applications in the North, is inevitably high in capital investment and low in labour intensity. It is big, not small; it is complex rather than simple. And the linkages it will make within a developing economy are likely to be limited. The trade-off between profit maximization and risk aversion is subjective, but it is very real for investors going into an uncertain situation, and it thus has very real technological implications.

Research and development provide an element of technology distortion more by their absence than their presence. In the mid-1990s, high-income economies were spending $218 per million people on R&D and the Asian tigers were spending about half that amount. Low-income countries, however, excluding China, were spending only one dollar per million people.[13] One of the most important areas for R&D in the South – investment in agricultural research – is at best patchy. Heavily dependent on donor funding and adversely affected by economic fluctuations, national research institutions have focused more on export crops than food, and less on food crops grown and consumed by the poor than on those consumed in cities. Members of the Consultative Group for International Agricultural Research (CGIAR), such as the International Rice Research Institute in the Philippines and the International Institute for Tropical Agriculture in Nigeria, devote considerable attention to appropriate crops, production systems and technologies. But appropriate technology is not their primary *raison d'être*, and with declines in ODA, their budgets have fallen under severe pressure. Some universities, such as Ghana's University of Science and Technology and the University of Zimbabwe, have dedicated technology institutions with a focus on appropriate and intermediate technologies, but in most Southern universities there is a lack of urgency and genuine practicality in research efforts.

Perhaps one of the greatest institutional distortions in the choice of technology can be found in the role played by multilateral and bilateral donor agencies – the very institutions most eager to remove distortions. In 1979, the ILO sponsored a study of donor influence in the choice of technology,[14] and in 1989 ITDG repeated the exercise with a questionnaire and detailed follow-up visits to some of the major donor agencies, including USAID, DFID, the World Bank, the Inter American Development Bank, DANIDA, NORAD and the Netherlands Foreign Ministry.

The change in the intervening years was profoundly discouraging. Although some agencies had technology advisers and vague technology guidelines, technology was, by and large, a non-issue. Donor agencies are prone to all of the engineering, economic and bureaucratic distortions mentioned above, but within a different context. One aspect of this context has to do with the broad interests of the donor country. For example, the British aid programme bears a distinct relationship to the evolution of the British shipbuilding, iron, steel, fertilizer and telecommunications industries. Canada's heavy assistance to the railway sectors of many developing countries is closely allied to official support for railway-related industries in Canada. DANIDA's strong support for the fishing and dairy sectors reflects obvious connections to some of the largest Danish industries.

Because donors have been heavily involved in certain sectors for many years, programmes become self-perpetuating. Such bias, while not inherently wrong, will push aid in a particular direction. As noted in Chapter III, for example, Canadian support for the transport sector in Bangladesh means railways;[15] for the Japanese it means bridges; for the British it may mean aircraft, while for other donors it means roads. What results is aid heavily skewed towards the mechanized sector, with none going to the non-mechanized sector which accounts both for twice the carrying capacity and for 60 per cent of all employment generated in transportation.

The immutability of some types of aid notwithstanding, there is an ebb and flow of fads among donor agencies which can be very costly, especially where technology is concerned. In the mid-1970s, the World Bank, for example, was very keen on appropriate technology. A thick 1976 report on the subject, one of many, stated that 'the Bank recognizes the importance of stimulating appropriate technology and the need to do more in this field ... It is expected that Bank support for appropriate technology will have a significant long-term impact on the choice and effectiveness of technological operations in the developing world.'[16] Over the next 15 years, the Bank, like UNDP, did a significant amount of work on the development of appropriate technologies in the water and sanitation sectors. But by 1990, the Bank's Science and Technology Unit had been dissolved and appropriate technology was more or less forgotten. As a general rule in Bank thinking at the start of the new century, and in line with the neo-classical thinking that underlies the issue of 'getting the prices right', technology is chosen according to least-cost criteria.

For most development agencies, the important technology-related issues at the beginning of the twenty-first century had not so much to do with how to promote appropriate technologies, as with how to limit the damaging effects of new technologies and their handmaidens: reductions in biodiversity, genetic engineering, tightening intellectual property rights, unregulated, instantaneous capital flow.

The latest fad – a sub-set of the globalization frenzies – might be called 'the knowledge boom', triggered by the revolution in communications technology. Noting the continuing problem that developing countries have in catching up with the rest, the World Bank says that now they:

> ... need not reinvent the wheel – or the computer, or the treatment for malaria. Rather than re-create existing knowledge, poorer countries have the option of acquiring and adapting much knowledge already available in the richer countries. With communications costs plummeting, transferring knowledge is cheaper than ever. Given these advances, the stage appears to be set for a rapid narrowing of knowledge gaps and a surge in economic growth and human well being.[17]

Compare that uplifting paragraph with the following: 'Since the discovery of Columbus, nothing has been done in any degree comparable to the vast enlargement which has thus been given to the sphere of human activity.' Or: 'It is impossible that old prejudices and hostilities should longer exist, while such an instrument has been created for the exchange of thought between all nations of the earth.' These two sentences, the first from *The Times* and the second from a book extolling a new technology as proof 'that nothing is impossible to man', were part of the celebratory hysteria that followed the completion of the Atlantic telegraph cable in 1858.[18]

The Victorians confused information with 'the exchange of thought'. Our era has assumed that its transfer – like the 'transfer of technology' – is a simple matter, in this case one of getting poor countries wired up to the Internet. In addition to forgetting that speedy information transfer is not new, we appear to have forgotten as well that speedy information has not always been used for good. The Bank might well ask, as it did in the sentence following the paragraph quoted above, 'Why then isn't this transfer [of knowledge] occurring as fast as we might expect?' One answer is that developing countries already have wheels, computers and treatments for malaria. They do not have to reinvent them, and mostly have not tried. Understanding new technologies, however, adapting them, paying for them and putting them to work in the cause of development is a different issue entirely. In South Africa, for example, arguably the best connected country on the continent where international communications technology is concerned, 75 per cent of the schools have no electricity and at universities there are as many as a thousand students for every computer terminal.[19] At the other end of the technology spectrum, American school boards are spending hundreds of millions of dollars every year on communications technology, on the assumption that computers will at last solve the teaching problem. They forget that by the turn of the century, some fifty million Americans had figured out how to use computers without any assistance whatsoever from the education system.

Even with better communications technology, information does not equal knowledge. Neil Postman defines knowledge as 'organized information – information that is embedded in some context; information that has a purpose, that leads one to seek further information in order to understand something about the world ... When one has knowledge, one knows how to make sense of information.'[20] The problem, he suggests, is not how to *move* information. 'We solved that problem long ago. The problem is how to transform information into knowledge, and how to transform knowledge into wisdom. If we can solve that problem, the rest will take care of itself.'

Not only is information not knowledge, information does not have to be factual or truthful or complete. An obvious example is the information favoured by development economists who focus, to the exclusion of many other things, on the market, high technology and macroeconomics. An indication of how macroeconomic issues have overshadowed what happens on the ground can be found in a comprehensive study of manufacturing in seven African countries. Although industrial development has always played a prominent role in African development thinking, and had a place in the World Bank's 1989 long-term perspective report, Roger Riddell finds that,

> There is, sadly, no evidence to suggest that these words will be translated in practice into effective, consistent and reoriented policy recommendations to African governments which the Bank is hoping to influence ... What is more, in its list of six key strategies for African development in the 1990s, the role and place of industry is simply not mentioned. Finally, industrial sector country studies produced by the World Bank in the late 1980s were still dominated by policy advice based on macroeconomic management issues, the overriding importance of market and price signals and the eclipsing of more interventionist strategies of industrial promotion.[21]

The same was true in the 1990s. A second donor-created distortion in technology lies in the business of tied aid. Much has been written about tied aid, and efforts have been made by some agencies to untie some amounts. By and large, however, about three-quarters of all bilateral assistance remains tied in one way or another, and a high proportion of the lending of international financial institutions is spent on hardware and services purchased in industrialized countries. Some of the distortions and disasters are well known, and have become public embarrassments. Britain's 1985 sale of 21 Westland W-30 helicopters to India, a deal concluded with great amounts of political arm-twisting, was essentially financed by deducting £65 million from the aid budget to India. A series of major engine malfunctions and one fatal crash led to all the helicopters being grounded within six years. A British Government-funded jute mill in Indonesia was somehow regarded as a 'technical success' despite a critical shortage of spare parts and a miscalculation of labour availability which kept its utilization at less than one third of capacity. Japanese bilateral assistance is managed by a collection of ministries, of which the Ministry of International Trade and Industry predominates. The result is a heavy emphasis on Japanese hardware in aid programmes, and a not uncoincidental subsequent commercial penetration in sectors opened by aid programmes. In 1983, an Italian contractor, funded through a $100 million EU contract, built a road from Mogadishu to the port of Kismaayo. By 1988 the road was impassable, even though the loan would not be repaid until 2023.[22] This put the EU somewhere in a queue with the World Bank, the IMF and other multilateral donors who were still recording Somalian debts of $558 million in 1999, hopeful perhaps that there might some day actually be a Somalia that could repay.[23]

A Canadian example, much beloved of investigative Canadian journalists, involved provision of a highly automated, Canadian-made bakery to Tanzania. Poorly designed, grossly overpriced, and purchased without consideration of other smaller, cheaper technologies already available in the region, the bakery became a beacon of inappropriate technology through tied aid. A 1983 study of Canadian aid to Tanzania observed that 'the economic costs to Tanzania and the domestic criticism CIDA has faced from media discussions should be sufficient to ensure that CIDA will not finance a similar mistake in Tanzania in the future.'[24]

Maybe not in Tanzania, but the feeble-minded donor approach to technology could be seen littering the floor of one of the GRATIS intermediate technology transfer units in 1991. Canada, long involved in rural development programmes in northern Ghana, required 350 handpumps over a five-year period for one of its projects. CIDA was already involved in supporting the development of a local pump-making capacity in Sri Lanka as well as in Togo, a few miles across the Ghanaian border. It was also, as the main supporter of GRATIS, actively involved in the development of a light-engineering capacity in Ghana. In reviewing the types of pump available, CIDA selected a design developed by the Swiss appropriate technology institution, SKAT. It was a model that could be repaired with relative simplicity by villagers, rather than professional technicians. So far, so appropriate. Then it was decided that the pumps had to be imported – from an

Alberta firm with no track record in pump manufacture. Over a quarter of the pumps arrived with components that did not fit together. In a kind of 'fitting' irony, the parts had to be sent to GRATIS – for *five* separate machining and strengthening operations.[25]

A further problem for developing countries in selecting distortion-proof technology is the paucity of information for good decision-making. Weak research institutions, inadequate libraries, small or nonexistent travel budgets and limited skill levels mean that decision-makers are often very limited in the choices apparently available, or that they become heavily reliant on consultants and aid agencies, who are prone to all the problems mentioned above. This is where the information superhighway can help. Five minutes on the Internet is enough to find basic information on the India Mark II pump, the Tara, the Afridev and the Shallowell pumps, with designs, prices and installation requirements. Whether more information would have been helpful in resisting CIDA's desire for a Canadian-made pump, however, is another question entirely.

Finally, there is the enormous distortion imposed by a country's power and influence structure. In an effort to maintain its support from key lobby groups within society (few of them representing the poor), a government may promote particular types of modernization, growth, research and development, with strong inherent biases towards the modern formal sector, and away from more appropriate types of investment. It may seem incredible that a bureaucrat or a politician would promote a smaller number of expensive houses, say, over cheaper shelter for more people – but where the poor have no voice, where the media are tame and parliaments weak, such decisions are far from uncommon.

Back to the future: policy options

Given donors' fixation with macroeconomic issues, and given their own prominent involvement in the transfer of inappropriate technologies, suggesting policy options for either donors or Southern governments is a bit like shouting into a hurricane. Perhaps the place to start is with the very idea of policy options. Despite the fashionable loathing for government meddling, much of this book has attempted to show that pragmatic government interventions, developed on a case-by-case approach, are both necessary and possible. This is *not* to suggest that 'getting the prices right' is a bad thing, but that getting the prices right, in itself, is not enough. As Lewis Mumford observed more than 60 years ago, 'the gains in technics are never registered automatically in society; they require equally adroit inventions and adaptations in politics.'[26]

Jeffrey Sachs, a more up-to-date Harvard economist, says that the international system fails to meet the scientific and technological needs of the world's poorest, in part because appropriate institutions like the World Health Organization are starved for funds, authority and even access to the negotiations between governments and international financial institutions where development strategies are negotiated. Investment in science and technology follows the market, eyes fixed firmly on scale, resulting in large, high-technology firms like Monsanto, Microsoft and Merck with interests in Northern problems and Northern markets. Serious investment in technologies that are appropriate to Southern problems is, Sachs believes, a responsibility of both the international aid system as well as Southern governments: 'Science requires a partnership between the public and private sectors. Free-market ideologues notwithstanding, there is scarcely one technology of significance that was not nurtured through public as well as private care.'[27]

Governments have, in fact, always been involved in technological change, from ancient times to modern, from industrialized Western nations to the newly industrialized countries of Asia. They compensate for market failure, they organize trade fairs and exhibitions, they create grant mechanisms, tax havens and research institutions. They create policy and they enforce it, and they also help in establishing society's attitudes towards innovators. The future of technological development in the South, especially in the poorest countries, will be determined to a large extent by the incentives and stimuli of the policy environment.

As noted in other chapters, education and credit can be important engines for poverty reduction, growth and the development of a technological capacity. In many countries these have been skewed, in access, volume and cost, in favour of the better off. They have been standardized for economies that will never be standard, and they have been designed to assist large-scale farmers and large-scale industry. Targeted interventions aimed at removing past distortions, and encouraging the development and adoption of appropriate technologies, are not inconsistent with more general reforms at a macroeconomic level. Nor are they historically inconsistent with the capacity-building technology interventions of governments in industrialized and newly industrialized countries.

Taxation and investment: a level playing field

Governments encourage the establishment of new business ventures, both local and foreign, through a variety of incentives which include special tax holidays, tax reductions and deductions, customs exemptions, guarantees, special permissions and protection from imports. Frances Stewart has pointed out that because governments tend to sweeten the package in accordance with the size of the investment, the playing field is usually tilted in favour of large-scale investors, and concomitantly more inappropriate technologies. Citing studies of industry in the Philippines and Thailand, she notes that firms benefiting from government promotion packages enjoyed a cost-reduction factor of between 35 and 39 per cent over non-promoted firms – all local and mostly small – and that semi-advanced technology had a further advantage over intermediate technology among those that were assisted.[28]

Some of the disadvantage suffered by micro-enterprise has to do with the definition of capital goods and the tax or duty attached to them. A lathe, clearly intended for industrial purposes, may be given preferential customs treatment, whereas tools used by micro-enterprises – hammers, planes, saws, sewing machines, bicycles, outboard motors – may be regarded as consumer items and taxed at a higher rate. Wittingly or unwittingly, the Philippines, for example, encouraged tractorization in the 1970s by eliminating all duties on tractors, but maintaining a 19 per cent duty on power tillers.[29]

Many governments have begun to place greater emphasis on small-scale enterprise development. Too often, however, this is a piecemeal effort, a sideline to the main objectives of industrial promotion. In countries where it is more than a sideline, the impact has been significant. In Costa Rica, for example, the government made a conscious decision, within the context of a severe adjustment programme, to foster and encourage both small-scale industries and the development of appropriate technologies. A comprehensive plan was based on the government's view that 'the market by itself does not assure that social objectives such as equity, reduction of extreme poverty, and increased social mobility are achieved.' Because of its job-creation potential and its linkages to other parts of the economy, small-scale enterprise was therefore placed on a more equal footing with large-scale enterprise, in part because of government's intention to 'support ... the process of adapting and generating appropriate technology to create a broad scientific base, with which Costa Rica can become a country that exports goods and services with a high technological content by the twenty-first century.'[30]

Technology assessment

The ability to assess technological alternatives is as important to government as it is to a corporation, or to a farmer attempting to decide between one seed and another. All technology leads to unforeseen results and complications; technology assessment is an attempt to minimize the negative effects of a particular choice. Technology assessment was transformed from an everyday event into a formal discipline in the late 1960s, when the United States Congress passed an act requiring an assessment of the environmental impact of those technologies and technological activities within the purview of the federal government.

Ernst Braun defines technology assessment as:

> an attempt to discover all the ramifications and effects which a technology is likely to have when it is in full use ... The study must be interdisciplinary in nature, requiring knowledge and insights from engineering and both the natural and social sciences. Both beneficial and harmful effects need to be described and alternative policies for dealing with this or rival technologies should be elaborated. Groups likely to be benefited or harmed should be identified and the study must be carried out impartially, both with respect to technologies and to social groups.[31]

This may seem a tall order, and as Braun notes, 'nobody pretends that is possible to foresee all the ramifications of a technology.' But without it, the haphazard hit-and-miss approach of the past is unlikely to change. A science and technology study of seven countries in Southern Africa found that virtually all production technologies were imported, many of them highly inappropriate to the needs they purported to address. Although all the countries had a variety of sectoral science and technology policies, these were fragmented and inadequate. Only two countries, Tanzania and Zambia, had a ministry of science and technology, and one of these was less than a year old. The minimal efforts at technology assessment in these and many other developing countries, and within most aid agencies, can only benefit from greater emphasis.

Laws and standards

A wide range of regulations and laws have important bearing on the choice of technology. Genuine interest in micro-enterprise development and the growth of technological capacity requires an *enabling* regulatory environment; one that establishes reasonable, enforceable standards, but which encourages rather than discourages talent and industry. The fact that the Kenya Standards Bureau approved fibre-concrete roofing tiles was one step in the product's advancement as a commercially viable roofing material. The next appropriate step might be the banning of the super-thin 19 mm corrugated galvanized iron sheets mentioned earlier, which absorb great amounts of foreign exchange and which rust within four years.

In China, the official recognition and approval of six different standards of cement made the spread of mini-cement plants much easier than in India, where only one standard was permitted. The comparative savings in investment, the increase in job-creation, and the growth in cement output were enormous. Encouragement is one thing, protection for successful investments is another. James Watt observed that 'an engineer's life without patents is not worth while.' Many appropriate technologies cannot be patented, but some can. The Botswana Technology Centre, for example, developed an inexpensive, battery-powered borehole water level sensor which had high export potential. Although patent protection for the sensor is available internationally, it was not, ironically, available in Botswana.

Many countries have enacted or inherited laws that are directly antagonistic to micro-enterprise development and appropriate technology. In Zimbabwe, for example, the Factories and Works Act imposes building, water, drainage and sewage regulations that are virtually impossible for small entrepreneurs to meet. Health, hygiene and safety regulations are enforced without adequate reference to the purposes for which they were established. In many countries, such laws are observed in the breach, but remain available as tools to be used against the informal sector when, on occasion, it overflows its geographic, economic or political boundaries.

In short, it is incumbent on governments to cut red tape. This means simplifying procedures, regulations and permissions. It means reconsidering such things as value-added taxes on small transactions, and rethinking labour regulations which, if observed, would simply put many small businesses out of business. It means revising laws and regulations that impede innovation in new and appropriate technologies; it means using incentives and disincentives to encourage appropriate technological change. And it means support for research and development into appropriate technologies, as well as the use of government procurement to encourage them.

A final note on regulations involves the question of non-tariff barriers to trade. Europe and North America have made a fine art of excluding competitive Southern products from their markets by establishing long lists of quotas, requirements, tests and standards. These are deliberate barriers that most Southern manufactures simply cannot surmount. Non-tariff barriers could perhaps be studied to advantage by developing countries seeking tools to fend off the inexorable southward flow of mass-produced Northern manufactures.

Attitudes

In attempting to determine why some societies have advanced technologically and others have not, technology historian Joel Mokyr places considerable emphasis on values, and a society's willingness to accept change. He observes that the Ancient Greeks valued sports and learning, while the Romans emphasized military and administrative ability. While both societies appreciated wealth, being brave or wise was equally, if not more important.

The exclusion of large parts of society from the development and benefit of these values – slaves, the poor, women – had, and continues to have a major limiting effect on their spread and general acceptance. Historically, the greater the chasm between the values of the educated, literate classes and the working portions of a population, the greater the negative impact on the development and diffusion of technology.[32]

At a more mundane level, it is hard to know what to say about the price of Ghanaian Coca-Cola being less desirable than that of the imported variety.[33] Regardless of what the World Bank and the IMF say about liberalization, there is a strong case to be made for taxing the advantage out of such a bizarre import. In the long run, however, only education, public awareness (including advertising) and a pride in things Ghanaian will take the place of tariffs. This should not be left to chance. Some very appropriate technologies have failed through inadequate promotion. The disastrous Canadian bakery in Tanzania might never have been built if the makers and promoters of smaller-scale bakeries had had their antennae up.

The importance of quality and of taste should not be diminished, however. Toyota, Nissan and Mitsubishi learned this lesson well, much to the chagrin of Chrysler, Ford and General Motors. To deny that the customer is at least sometimes right is to deny history. Too often what is labelled an 'appropriate technology' is poorly finished or unprofessionally packaged. Some appropriate technology is labelled second-class because it looks second-class. And some is labelled third-class because it does not work. Had there been time and assistance for Ghana's small-scale soapmakers, they might have been able to add inexpensive perfume and an attractive wrapping to their product before Lever Brothers drove them out of business. Professional finishing and competent marketing are essential ingredients of successful technology development.

Institutions

Institutions have played an important part in the development and transfer of technology throughout history. Because the poor are under-represented in places where decisions are made, however, they are simply not heard, and an appropriate technology that might benefit them often receives little attention. Special efforts are required to ensure that small producers have adequate access to credit, raw materials, agricultural inputs and foreign exchange. In Zimbabwe in the early 1990s, tin shortages were particularly hard on informal-sector tinsmiths, for whom tin was their entire livelihood. Because of the political strength of formal-sector manufacturers, however, small-scale users were ignored. A similar situation prevailed in Ghana in the mid-1980s, with regard to foreign exchange. At the time, an auction system was in use, but there were minimum amounts for which one could bid. Even if small entrepreneurs – in need of something as basic as welding rods, available across the border in Togo – knew how the system operated, their needs were so small that they were simply excluded from the system.

NGOs, which have broad mandates and better resources than many Southern research institutions, and more practical output, are often less disciplined. They have weaker institutional memories, and are less able (or less willing) to work with governments and the private sector. NGOs may work towards 'empowerment', but most have serious problems of scale. NGOs are favourably disposed towards the concept of appropriate technology, but with notable exceptions, most have very little technology awareness. Somehow this must change.

Co-ordination seems an obvious solution, but 'co-ordination' is another of those words so frequently used among development agencies that its meaning has all but evaporated in hot air. To suggest that better co-ordination is required between research organizations, educational institutions, government agencies and departments, donors, NGOs and the private sector is to state something so seemingly obvious, yet so seldom done, that the idea may be a waste of time. Because organizations are so loth to share and to listen, many genuine attempts at co-ordination wind up placing more emphasis on the process of co-ordinating than on the intended result.

There are, however, glimmers of hope on which interested groups and institutions could build. The United Nations Development Fund for Women, UNIFEM, for example, works in association with other United Nations agencies, but is also closely allied with a wide range of research, bilateral and non-governmental organizations, with a primary programming emphasis on small-enterprise and informal-sector development, appropriate technology and innovative credit programmes for poor women.

Another example can be found in Botswana, where a study recognized the inadequacy and costliness of the country's fragmented approach to science and technology, but saw the weakness in other countries' attempts to create formal research or policy institutions.[34] The study acknowledged that Botswana did not yet have 'the capacity to effectively identify, acquire, adapt, develop, service and use science and technology', and that a policy was required which could 'guide national action for technological transformation, and in particular to reduce the country's state of technological dependency.' The study saw needs in several areas: human resource development, economy and production, the need for consistent and appropriate standards and legislation, research and development, information and public awareness, and institutional development. It also recognized that there were greater or lesser degrees of expertise in some or all of these areas, and proposed a short-, medium- and long-range plan. Rather than create yet more institutions, the plan aimed to bring key officials from government, the private and non-governmental sectors, and the educational community together in a variety of co-ordinated fora for the development, co-ordination and control of a thoughtful and genuinely national policy on science and technology.

Without such a policy, a developing country is left with a loose set of funding programmes, *ad hoc* co-operation between universities, industry and government, and a lot of windy rhetoric. The development of a technological capacity and of technologies appropriate to the needs of any country is dependent upon the institutional, economic and social environment in which they grow. For technology to advance, a government needs good information on the scientific and technological aspects of its economy and infrastructure. Policy-makers, administrators and the private sector must be able to recognize and act upon the technological aspects of their work. This requires training, co-ordination, the sharing of information. It means that when opportunities and problems arise, genuine alternatives are considered and weighed openly. It means that there must be good interaction between engineers, economic planners, entrepreneurs and social anthropologists. Such factors are the hallmarks of any society in which technology is effectively harnessed as a tool for development. In countries where technology has not been institutionalized, but where it is nevertheless an essential key to the reduction and elimination of poverty, they are even more critical.

CHAPTER XVI
Mastering the machine

These are the days of miracle and wonder; this is the long-distance call.
Paul Simon

IT IS SOMETIMES said that the world is engaged in a war, and that the enemy is poverty, an enemy that robs adults of their youth and children of their lives. This type of thinking, vaguely reminiscent of Lyndon Johnson's 1964 'war on poverty', is a conveniently innocuous way of depersonalizing 'the enemy' and of avoiding the wrath of real people and real institutions. It is a bit like saying that in a war, the enemy is dead and wounded soldiers. It avoids the unpleasant idea that both rich countries and poor countries, their politicians, their governments, their village elders, their aid agencies and their vested interests may all be a part of the problem – part of 'the enemy'. It avoids confronting the possibility that the war is not against poverty, but against the short-sightedness and selfishness that spends billions of dollars on cigarettes, golf courses and a high-technology war in Kuwait and Kosovo, but spends virtually nothing to prevent the slide of other countries – many of them in Africa – into anarchy and chaos, virtually nothing to educate young girls in the South, and only a pittance on the health care that could save millions of lives every year. Each day, 33 000 children die of ordinary malnutrition and preventable disease, more in a year than all the people who live in Belgium.[1] Hundreds of millions of individuals live in squalor, illness and hopelessness, on the edge of an economic abyss. Poverty – the product of inaction, entrenched vested interests and short-term greed – saps people and nations of their vitality and their dignity; it creates political instability, exacerbates ethnic tension and contributes to irreversible environmental damage. As a new millennium begins, there can be no more important international objective than to deal effectively and decisively with this scourge of the last one.

Yet viewed from the trenches, the so-called war against poverty seems half-hearted, prosecuted with far less vigour and money than went into the

West's military confrontations with Iraq and Serbia. So incomplete is our knowledge of the conflict, so careless our approach and so weak our resolve, that we are unsure if the number of absolute poor is a billion, or half again that number. Our generals have little direct contact with the enemy as identified, and rarely visit the forward command posts. Skirmishes and pitched battles take place along a front which changes from time to time, but there are few meaningful advances. Sometimes the generals bring up heavy artillery and mass troops against what they think is a soft point. And sometimes they succeed in breaking through enemy lines. The eradication of smallpox, for example, was a battle that ended in a spectacular victory. The multi-billion dollar investment in integrated rural development, however, left the land barren and the front unaltered. In some cases, as with the Asian economic meltdown or civil strife in Africa, the alleged enemy breaches poorly defended barricades. The result is increased environmental damage, refugees, chaos, government breakdown, and failure of the rule of law.

After years of low-level proxy warfare, both generals and public are shell-shocked, inured to battle. Part of the problem is that we have been conditioned to believe in the 'big breakthrough' – a cure for cancer or AIDS; the miracle food, the miracle drug, the miracle economic solution. As Michael Lipton observes, 'Scientists, like economists, prefer *interesting* problems to *important* problems.' This explains why some donors have 'found the solution' to poverty in micro-credit, and the solution to credit in minimalism. Others have found it in 'getting the prices right' and the quick fix of the market. Beguiled by a misplaced belief in the big breakthrough, independent African governments began life by spending vast resources on what they hoped would be a great leap forward into industrialization, at the precise historical moment when the prices for their exports began to fall, passing the price of oil on its way up. Today, new solutions are touted in the 'knowledge revolution', 'warp speed' communications and ever-higher technologies, catching paralysed generals like deer in the on-coming headlights of the thing called globalization.

There is, unfortunately, no miracle drug; no sudden cure for cancer or AIDS. In health, as in all fields of human endeavour, advances are made by individuals working long and hard, for years, in dozens of countries, barking up wrong tree after wrong tree until – occasionally – they find the right one. Sometimes their hard work is lost – through accident or malice. Sometimes it pays off through sheer good luck. There is a story about Norman Borlaug beginning his first trials in the 1950s with a cross-bred dwarf wheat in Mexico, the wheat that led directly to the start of the Asian Green Revolution only five years later. What subsequently became known as a 'miracle' wheat was hit during that first planting by a rust, and the entire crop was lost. Distraught, Borlaug checked the seed bags, and to his great relief, found a few kernels in one. The descendants of the kernels in that bag are being planted today by millions of farmers around the world.

Humans love drama; we enjoy tales of discovery, and so the 'big breakthrough' version of history has special appeal. Brought up on a cinematic diet of Mickey Rooney as young Tom Edison, Ralph Richardson as Frank Whittle inventing the jet engine, and Curt Jurgens as Wernher von Braun, Europeans and North Americans are especially impatient for quick results. Their comic-book version of science and the rewards of hard work is reinforced by life in a world of seemingly quick results, a world of ever-changing, ever-improving consumer goods – computers, video recorders, the Internet, silicon chips, CD players and microwave ovens. With so much progress around them, it is little wonder that they become impatient with decades of promises from agencies such as Oxfam, CARE and UNICEF, promises clearly at odds with the consistently negative daily headlines from the South.

There is another reason for public impatience, and it lies at the heart of the way the development community conducts its affairs. Bilateral donor agencies couch budgetary requests to their parliaments in terms of the impact of their work on the reduction of poverty. That great slices of their budgets never actually leave the home country, that much is spent on inappropriate goods and services as an expensive means of achieving limited commercial and political objectives, that an infinitesimally small portion is actually devoted to direct poverty-reduction programmes, is seldom mentioned. When the results show that 'developing' countries have slipped ever deeper into poverty, legislators, journalists and the public become understandably suspicious about whether the money is actually getting to those who need it most.

Fuelled by the hypocrisy of donor agencies, public cynicism is stoked by the image-making and the unrealistic claims of many non-governmental organizations. Their most striking appeals are for disaster relief, which often contributes to an image of hopelessness in the South. Their general fundraising appeals inevitably suggest that development is cheap, simple and quick. This approach has encouraged a lot of donations over the years, but its appeal is wearing thin. There is talk of 'aid fatigue' among the public. Perhaps the issue, however, is not aid fatigue, but fatigue with aid agencies. Too many agencies; too many promises; no explanation for continuing misery.

The cheap-simple-and-quick approach to fundraising has spilled over into the professional arena as well, with NGOs making their proposals to bilateral agencies and foundations in precisely the same way. In order to 'sell' a project, unrealistic claims are made about what can be achieved with limited funds and low administrative overheads, usually in only two or three years. High expectations are raised, which must then – only a year or so later, when more funding is required – be supported by positive results. Failure (or more likely, in the short space of time available, lack of demonstrable success) is therefore concealed for fear of jeopardizing funding. Successes are exaggerated accordingly. As publicity departments become more fulsomely positive (with limited reference to past errors, now happily corrected), independent reporting on the aid business becomes increasingly negative. In *Lords of Poverty*, a popular and very negative book on development assistance, Graham Hancock begins with a dedication to senior World Bank staff who obtained an early synopsis, and who attempted to limit his access to information. Their behaviour convinced him that 'the aid business does indeed have much to hide.'[1]

When the means bear little direct relationship to the ends, when targets are unduly exaggerated, and when further funding is directly related to the achievement of impossible goals, a premium on dishonest reporting is the natural outcome. Worse, after two or three decades of it, aid workers have begun to believe their own publicity material. Governments resist reports of failure; civil servants fall deaf and mute; NGO promises escalate. But development is not a publicity challenge. It is a real test of real skills, ingenuity and commitment. It is the most crucial test facing this generation and will probably be the most crucial issue for the next two or three.

In some ways, the appropriate technology movement has been fortunate in remaining on the fringe of 'mainstream development thinking'. This has happened for two reasons. The first is that Fritz Schumacher said it would be straightforward, and it was not. He said that if you got the technology right, everything else would fall into place. But it didn't. The second is that appropriate technologists sometimes have difficulty pointing to 'big success stories'. There are, of course, many – some described in this book. Hundreds of thousands of increasingly inexpensive handpumps have made life infinitely better for poor people in dozens of countries. Similar numbers of cheap latrines have prevented disease, and millions of lives have been saved through the development of oral rehydration therapy. Better living conditions have ensued, and millions of jobs have been created in the manufacture and construction of these technologies, and in the production of fuel-efficient stoves, or in the increased output of high-yielding varieties of wheat and rice. Important contributions have been made through the production and availability of low-cost cement, cheap electricity, inexpensive, durable building materials; in the production of small farm implements and windmills. The creation and maintenance of tens of thousands of miles of hand-made rural roads has created jobs and improved living standards for entire communities. Success can be found both in the large numbers like these, and in the single application. An inexpensive loom can make all the difference between a family's poverty and prosperity. The same is true of improved potters' wheels, of new tools for carpenters and blacksmiths, better ovens for bakers, and boats for fishing families.

But while these achievements, big or otherwise, are important, in a sense the small incremental achievements that eventually made them possible are even more important. The six small iron foundries established by GRATIS and the Technology Consultancy Centre in Ghana between 1986 and 1991 created perhaps two dozen jobs and turned out a lot of corn-mill grinding plates that could have been bought in Europe for a few thousand dollars. But those foundries were probably more important to the future of Ghana's industrial development than almost any other technology imported or developed there in the previous twenty years. For the first time they opened a door to the local design and manufacture of machines, a fundamental advance on repairs and simple metal fabrication.

Adequate time for product development and diffusion is also a critical element. It took centuries for the heavy plough to become widely accepted in Europe. It was a century and a half after the first Montgolfier balloon flights that commercial air travel became a real possibility. Without five years of development work, the humble fuel-efficient Sri Lankan stove would never have made it into the market-place. Without ten years of dogged persistence and the constant refinements of John Parry, fibre-concrete roofing tiles might, after the initial experiments, have become just another failed effort.

Technique – invisible in the annual report of an appropriate technology organization – is often more important than technology. A small change in the way a bolt is made, for example, can make the difference between the possible and the impossible, the economic and the uneconomic.

If the invention of the steamboat had been an aid project, we might still be hoisting sails. The failed experiments of Robert Fitch and others would have ensured the cancellation of funding, and fifteen years later Robert Fulton might still have been in the rope-making business where he began. On the other hand, something entirely different might have happened. Chapter III lamented the complete lack of donor interest in Bangladesh's non-motorized transportation sector. Faced with increasing competition from the growing number of mechanized river craft, a few country boat-owners began to experiment in the mid-1980s with small 12 to 16 horse-power diesel irrigation pump engines – normally not used during the monsoon. With a few modifications and the addition of a shaft and propeller, they found that sails were no longer necessary. A simultaneous reduction in import duties on pump engines helped fuel a transportation revolution unlike anything seen in the Ganges-Brahmaputra delta in a century. Within five years, 75 per cent of the country boats – thousands of them – had been mechanized. Without a penny from an aid agency, one of the last great survivors of the age of sail had passed into history.[2]

What a tangled web: the importance of failure

There is no shame in honest efforts that fail. Failure is a common feature of business the world over. Product failure is known to every manufacturer in every country. Among car makers, the Ford Motor Company's 'Edsel' is synonymous with failure. Charles Goodyear, who invented the vulcanization process for rubber in 1839, died a pauper. Leonardo Da Vinci, one of the greatest technologists of his era, never published a word of his thousands of pages of engineering notes. And everyone who works in the field of international development lives with failure on a daily basis. Yet because of the premium that has been placed on dishonesty, these failures are rarely acknowledged and seldom studied. The lessons they have to teach remain hidden, and the inability to admit failure breeds a pathological inability to learn from failure.

This is especially true of investments in social development, where returns are difficult to calculate and weaknesses are correspondingly easier to obscure. It is also true of large-scale infrastructural development, not because returns are difficult to calculate, but because the final net return is often not apparent for years, until everyone associated with the project is safely ensconced elsewhere. The impassable road that Somalians might have paid for until 2023, had their country not disappeared, is one example of this. Another is the multi-billion-dollar investments made in integrated rural development in the 1970s. Many of those projects, which included expensive foreign advisers, and which were covered by loans that children not yet born will have to pay off, were abandoned not because they failed. They failed – at least in part – because they were abandoned. Many could have been salvaged if donor agencies and governments had been willing to learn the lessons they had to teach, and if they had been willing to wait for the lessons to become clear. But they were abandoned because the development wheel had turned another quarter revolution; new planners and new economists had new and better ideas about development.

Wheels are thus reinvented time after time, until wheel invention becomes a virtue in itself. Insiders fail to question the wastefulness of demonstration projects that demonstrate nothing. Monitoring reports tell us whether a wheel is a good one, not that the wheel was not required in the first place. Evaluations tell us that the locomotives sent to Bangladesh work or do not work; but do not say if locomotives were a good or bad idea in the first place. Given the size of the problem development agencies tackle, and given the state of the world, the elusiveness of success should not be a surprise. Nor is failure really a secret. But treated as a secret by aid agencies, failure attracts the curiosity of reporters. And the most zealous reporters, taking publicity material produced for taxpayers at face value, naturally produce sensational stories, adding to the tangled web of public misunderstanding.

Southern governments

Southern governments, of course, bear the ultimate responsibility for what happens within the borders of their own countries, whether they are able to control events, or not. This book has made the point that much of what happens in the South lies beyond the control of Southern governments. This is not to suggest, however, that they are helpless, or that they should be absolved from responsibility for many of the errors of the past. Governments have not always worked for the common good. Too often they promote the interests of special groups: the armed forces, landowners, the private sector, urban interests over the rural. Inefficiency and poor financial performance have persisted. Many countries have promoted inappropriate industry and neglected agriculture, looking inward rather than to regional or global opportunities. Limited as the options may have been, even the existing few have often been ignored.

Although perhaps over-emphasized in Northern thinking, corruption and 'rent seeking' in the South cannot be dismissed as irrelevant or inevitable. Democracy is another thorny issue. Responsible government has proved remarkably durable in many countries – Jamaica, India, Senegal and Botswana, to name four. In others, such as Peru, Thailand, Brazil and Chile, it repeatedly survives the generals who interfere with it. But in too many countries, including some of the 'newly industrialized countries', political repression, one-party governments and military regimes have been the order of the day. Corrupt and authoritarian regimes negotiated many of the loans that triggered the debt crises of the 1980s and 1990s.

Despite an understandable Southern resistance to 'universal' values that appear to originate in the North, democracy, justice, honesty and respect for human rights are not concepts alien to the South. Nor are they things that can be imposed from outside. But they can be encouraged and supported through assistance to social movements and non-governmental organizations, and through financial and technical support to other pluralistic institutions. With stronger mandates and better support, the United Nations, regional organizations, and non-governmental bodies such as Amnesty International and the International Commission of Jurists could play more prominent and effective roles in monitoring elections, and in monitoring the observance of human rights and democratic institutions. Recognition of the importance of these things, and of where ultimate responsibility lies, however, is clear. It was expressed succinctly in the final paragraph of the 1990 Report of the South Commission:

> In the final analysis, the South's plea for justice, equity, and democracy in the global society cannot be dissociated from its pursuit of these goals within its own societies. Commitment to democratic values, respect for fundamental rights – particularly the right to dissent – fair treatment for minorities, concern for the poor and underprivileged, probity in public life, willingness to settle disputes without recourse to war – all these cannot but influence world opinion and increase the South's chances of securing a new world order.[3]

Donor organizations

In the absence of fair, intelligent and consistent approaches to the issue of technology by large bilateral and multilateral donor agencies, and in light of the single-minded enthusiasm with which they pursue market-dominated policy lending, despair might seem the most logical response. Policy looms large in donor thinking, not so much because of structural disequilibria, but because of project fatigue within donor organizations themselves, and an exaggerated sense that 'their' projects have failed. The logical (defensive) next step is to blame failure on wrong (recipient) macro policies rather than (donor) project design. Add to this the toxic idea that donor loans and conditions can push recipients to 'get the policies right' (just as the now forgotten loans for integrated rural development were supposed to get rural development right), and the magic bullets of structural adjustment and growth-oriented, market-driven solutions are at least partly explained.

There is a place for adjustment lending and programme support, but it should not take the place of carefully developed, *technically sound* projects. For those donors unable or unwilling to make changes in their own policies, but who are interested in technology development, it may make sense to simply focus on a few critical areas in which appropriate technology may be advanced. Well-aimed rifle shots are more likely to strike a target than a handful of bullets tossed into a campfire. The function-based approach, discussed in Chapter XIII, is one kind of rifle shot. Work on micro- and small-scale enterprise development is an example, as is the dedicated focus on credit by Grameen Bank and a dozen similar organizations in Bangladesh and elsewhere.

The sectoral approach is a different kind of rifle shot. The sectoral approach focuses comprehensively on a key area – such as silk production in Bangladesh – which can involve large numbers of poor people and which can attract government attention at a policy level. Chapter X suggested another sector: light engineering and the manufacture of machines – oil presses, wood-turning lathes, brick moulds, corn-milling machines, rice huskers, hammer mills and the like. Historically, machine design, fabrication and adaptation have been a critical part of the evolution of mechanical industries in the North. And as the Ghanaian examples show, light engineering can develop key linkages with the more modern, formal sector as well as with agriculture. For a donor organization, there is a further advantage. By focusing on machine *makers* rather than machine *users*, the cross-sectoral and multiplier effect of the intervention can increase exponentially.

Non-governmental organizations

Alone, most large donor organizations are unsuited to labour-intensive small-scale interventions that require programming flexibility and time. If poverty reduction is to be given greater prominence in their spending, if more appropriate technologies are to be developed and used, new and improved programming mechanisms will be required. One of these is support to non-governmental organizations. NGOs are a key source of research and development for appropriate technologies, techniques and approaches. Some of the most promising experiments and examples of 'scaling up' have been conducted by organizations like BRAC in Bangladesh, the Self Employed Women's Association of India, Grameen Bank, CARE, World Vision and Oxfam. However, there is a tremendous amount of wasteful duplication and amateurism among NGOs. There is much competition for funding and 'territory', a lack of cohesion, and poor institutional memory – all of which contribute to intellectual sloppiness and the ritual re-invention of wheels. This is particularly and increasingly true of Northern NGOs, which are in danger of losing their *raison d'être* as their developmental legitimacy is appropriated by their Southern counterparts, and as bilateral organizations encroach on their funding role.

Those non-governmental organizations that can overcome their traditional abhorrence of the profit motive have an increasingly important role to play in fostering micro and small enterprise (as opposed to the simple provision of credit). Profit – the surplus of income over expenditure – is the vulgar word for what most NGOs have practised for years under the polite term, 'income generation'. In an unfavourable climate for micro-entrepreneurs and in situations of market failure, NGOs have often been the innovators, the entrepreneurs, the carriers and developers of technology. They have struggled against almost insurmountable political odds and have been used, abused, ignored and then lionized by the official aid community, depending on the fashion of the day. But in many of the poorest countries, something else is going on. The climate for micro-enterprise development has improved, and the informal sector is receiving more attention and support. NGOs must therefore be prepared to change roles, to become facilitators for real entrepreneurs, rather than to act as entrepreneurs themselves. With care, and with the development of strategic alliances among themselves, with small-scale entrepreneurs and perhaps with parts of the academic and research communities, the investments they have already made stand to pay large development dividends.

BRAC's creation of a large-scale self-financing credit operation for the poor is one example of how some NGOs are becoming increasingly professional. Another is taking place in Bolivia, where the non-profit development foundation, PRODEM, helped to form the first commercial bank in Latin America – Bancosol – dedicated to micro-enterprise development in urban areas. PRODEM has continued its work in rural areas and in 1997 had 36 000 clients and an active loan portfolio of $18 million.

Another area of vital importance is the environment. Northern environmental agencies have been very successful in getting environmental issues onto the agenda of development organizations. But the reverse – getting poverty reduction and development onto the agenda of environmental organizations – has not happened to any great extent. Individually, development and environmental agencies may each be able to point to success stories. Together, forming strategic alliances and using their comparative advantages, they could be considerably more effective. Bringing government into the mix can add even greater value. An example is Pakistan's National Conservation Strategy (NCS). The NCS was drafted at the height of environmentalist optimism in the early 1990s by a coalition of CIDA-funded Pakistani environmental and development NGOs, working with government officials. By the end of the decade, the blush was off the environmental rose, but the same coalition – seasoned by the harsh economic and political realities of Pakistan – was still at work, drafting new legislation, working with journalists and lawyers, and taking a longer view of building a public constituency for environmental protection. More importantly, by working consistently with key government officials, the NGOs had created important space that had previously not existed for environmental and development policy dialogue.

More and more Southern NGOs, like BRAC, PRODEM and those in Pakistan, are moving forward into new programming areas – geographical, social, technical or policy-oriented. If they are to succeed, they will need assistance, and they will need greater recognition. Sustained financial assistance is one thing; few Southern NGOs have a large or secure domestic funding base. Good publicity is another. But genuine political recognition and support is a different matter entirely. Despite the fear some governments have of NGOs, strong, pluralistic societies need the challenge, the alternatives and the debate that a vibrant civil society can provide. This is as true in the North as it is in the South, perhaps even more so where international development is concerned.

As Southern organizations grow and mature, foreign NGOs – something of an aberrant link between one society's concern and another's need – will have to adapt or they will die. The financial and moral support that some offer will continue to be important to their Southern counterparts, but that will not be enough. For many Southern NGOs, the old connections no longer play a major financial or functional part in the relationship. Alliances must be rekindled, therefore, along more strategic lines. Survivors among Northern NGOs will have a genuine speciality to offer – technology services, specialized credit, health or educational knowledge. But the most important feature of North–South NGO partnerships will be alliances formed around critical policy issues: thoughtful, balanced, professional policy interventions on issues in health or education or the environment; and in broader areas such as trade and investment policies, and aid policy. And it will be found in the development of strong, pro-poor public opinion in the North and the South, opinion so strong and so vocal that it cannot be ignored any longer by governments.

People

Although this book has focused mainly on technology that is small, simple, cheap and non-violent, it is not intended as an attack on higher technologies. Many apparently simple technologies are derived from intensive, high-technology research and development. The bicycle, for example, which was not successfully commercialized until well after the typewriter and the sewing machine, combines specialized techniques of thin-walled drawn-steel tubing, modern rubber-processing technologies, ball bearings, sprockets, roller chains and complex gear-changing mechanisms. The issue for developing countries is not a trade-off between 'high' and 'low' technologies. It is a trade-off between appropriate and inappropriate technologies.

One of the lessons of development, much acknowledged but less practised, has to do with participation. Effective appropriate technologists have learned the hard way that the uptake and sustainability of new ideas and new technologies require the involvement of the intended beneficiaries from the outset. Building their capacity to manage change is as important as making a better mousetrap; in fact it is a fundamental prerequisite to ensuring that the mousetrap is better, and is actually used. As Schumacher put it, 'Find out what people do and help them do it better.' Help them do it better. Much of the technology problem has to do with people; not 'dirty fingernail people' or entrepreneurs, but with those in government, large corporations and donor agencies who control budgets and ideas – 'bureaucratic man', 'economic man' and 'engineering man'.

Robert Chambers' well-known book, *Rural Development; Putting the Last First*, is not really about rural development, it is about the scope for personal choice among those whose work has an impact on the lives of poor people; it is a plea for development workers to 'exercise imagination in thinking through the distant but real effects on the rural poor of technical and policy decisions, and of outsiders' actions and non-actions.'[4] It calls for a better understanding of two things commonly discussed among donor agencies and NGOs, but not always well understood: the first is groups, and the second is individuals. Groups are an important vehicle for reducing the risk element in new ventures, for the development of economic and political empowerment, and for the nurturing of participatory development – something that in itself is often misunderstood. Whether participatory development is an essential part of a project, or an end in itself – a necessary prerequisite to socio-political and economic development – is an old debate. The fact of its importance is not an issue. But what is sometimes forgotten about groups is that they are made up of individuals, real people with different perceptions, different memories, different strengths and different needs.

It is ironic that donor agencies which place increasing emphasis on the individual at one level, work against individuality at others. The growing interest in micro-enterprise development and rural credit programmes is the positive side of this conundrum. But on the negative side is the fact that individuals are often ignored. As this book has attempted to demonstrate, many of the most successful aid projects flow from the skill, tenacity and longevity of very special individuals. This should not be surprising, because the same is true of successful businesses and successful organizations everywhere. Development literature and case studies ring with the accomplishments of organizations like Grameen Bank, Sarvodaya, SEWA and BRAC, but oddly, beyond a limited conference circuit, the key individuals that created these organizations are unknown. A search through the indexes of books on development will turn up multiple references to academics such as Sen, Chambers and Streeten – and so they should. But there is virtually no mention of Akhtar Hameed Khan, Hector Morales, Ela Bhatt or F.H. Abed. It is as though the remarkable organizations which these social entrepreneurs created, and which bear the unmistakable stamp of their own personalities, had appeared full-blown, part of a 'phenomenon' rather than the work of especially committed, competent individuals.

The failure to recognize this stems in part from the way donor organizations themselves are organized and managed. They are reverse images of almost everything business writers find in surveys of successful companies.

They are secretive, compartmentalized and highly centralized in their decision-making structures. In an inherently risky business, they avoid risk. Individuals seldom remain in the same job, or even the same country programme for more than two or three years. Overseas assignments are generally two years in length, less than the gestation period for the average bilateral aid project. There is little tolerance of failure, few opportunities for constructive debate, and a reward system which encourages control, tidiness and the smooth flow of paper. Large numbers of men and women in big donor agencies are genuinely committed, hard-working individuals, but within such a system, it is not surprising that commitment, risk-taking and debate can fall early victims to the need for a successful career.

Foresight as hindsight

In 1863, the young Jules Verne produced a futuristic novel, *Paris in the Twentieth Century*. Set in 1962, it foresaw a world lighted by electricity. There were horseless carriages powered by internal combustion engines, streetcars and driverless trains, elevators, automatic doors, copiers, calculators and fax machines. Verne, however, was building on what he could see – prototypes or working models of all these things were actually in existence by 1863. In Verne's 1962, however, there was no income tax, no typewriters or steel-nibbed pens; clerks still wrote with quills. Even impoverished students still had servants. War, disease and newspapers had vanished.[5]

In 1911, the science editor of the *New York Times* speculated on 'the flying machine of the future'. It would be 'a kind of air-yacht, weighing at least three tons . . . It will carry several persons conveniently, and . . . it will fly at speeds of 150 to 200 miles an hour'.[6] Writing in 1938 about the world of 1958, Arthur Train Jr. predicted a great deal of television, devices that resembled cell phones and personal computers, and the convenience of the morning paper arriving by fax machine every day. Cars would be used mostly by children, furniture would be made of magnesium alloys and beryllium (whatever that is), and 'occasionally plastic'. In addition to automatic garage door openers, there would also be automatic window and refrigerator door openers.[7]

The future, of course – as Neil Postman observes – is an illusion. 'Nothing has happened there yet.'[8] That is why so many predictions about the future are based on the past or the present. What then are we to say about the future of poverty, aid and technology? Where aid is concerned – the business of giving and receiving – John Steinbeck wrote:

> Perhaps the most overrated virtue on our list of shoddy virtues is that of giving. Giving builds up the ego of the giver, makes him superior and higher and larger than the receiver. Nearly always, giving is a selfish pleasure, and in many cases it is a downright

destructive and evil thing. One has only to remember some of our wolfish financiers who spend two-thirds of their lives clawing fortunes out of the guts of society and the latter third pushing it back . . .

It is so easy to give, so exquisitely rewarding. Receiving, on the other hand, if it be done well, requires a fine balance of self-knowledge and kindness. It requires humility and tact and great understanding of relationships. In receiving you cannot appear, even to yourself, better or stronger or wiser than the giver, although you must be wiser to do it well.[9]

Poverty has reached such an unacceptable world level, not because the world has no tools to deal with it, not because we are short of ideas, but because we are short of the political will to use the tools properly and to implement the ideas effectively. 'Lack of political will,' however, is a convenient term for summing up the shortcomings of others. It usually means that the rich and powerful are acting in their own best interests, but it is a catch-phrase that stops short of discussing how those interests – local and international – capture and distort international development efforts.

This book has attempted to answer some of these questions. It suggests that aid as presently conceived and practised, is largely irrelevant to the needs of poor people. The growth principle, on which much aid spending is predicated and justified, is important, but it has severe limitations. Greater efficiencies in the use of what we have and what we know could provide much better long-term benefits. In any case, it is by no means certain that much aid actually contributes to growth. And if it does, it is by no means clear that growth is reaching the poorest in a positive way.

One of the solutions to the poverty problem, an important one, is technology. Unfortunately, in the hands of governments, foreign engineers, experts, amateurs, bureaucrats, corporations and aid officials, the lessons of technology in history have been largely ignored. In the quest for sales and a search for the big breakthrough, they have colluded in one of the most expensive and tragic hoaxes of all time. Vast numbers of inappropriate tractors and threshers and factories have been shipped south, at ever-increasing prices, to be used at a fraction of capacity or to end their days prematurely, rusting in the rain.

Today's poverty, worse than at any time in history, is, as Schumacher said, a monstrous and scandalous thing, totally abnormal. And there is no reason, despite brave donor agency calculations, to think that the situation is being reversed. The development community has, in a sense, stopped rearranging the proverbial Titanic deck chairs, because the ship's list is so great that all the deck chairs have shifted to the right. The band plays on, while passengers on the upper deck take comfort in singing hymns to market economics, keeping one hopeful eye on the lifeboat called 'sustainability' and the other on the potential for some last-minute sales. But the davits have become rusty over time, and the cables are snarled. We forget that 'sustainability' is relative. It cannot be packaged up into two-year projects. Sustainability of the global environment, or what is left of it, is not compatible with current levels of poverty. This is true at a local level – whether in Turkana or what remains of the Malaysian rainforest – and it is true globally. 'Earth days', 'earth summits', green political parties and environmental checklists in aid proposals are all useful, but they are not enough, and they will amount to very little if the fundamental issue of poverty is not squarely faced.

The Titanic provides a glib analogy, but the great ship and some of its modern-day parallels are also real-life lessons in technological hubris, showing the error in assuming that bigger, faster and more complex is always better, more appropriate and more sustainable. There is another wrong assumption: that the world can move forward into the bright sunlit uplands of peace and development when one set of people has the ability to harness technology for production and social benefit, while another set of people lacks this capacity, and is dependent on the former for scraps from the table. There is yet another odd assumption: that wealth and privilege can survive amidst growing poverty and political mismanagement. The bloodiest revolutions in history have proven this assumption to be wrong.

At a more mundane level, we have forgotten that originally, as set out in the Pearson Report, structural adjustment was a concept which involved industrialized countries adjusting to greater exports from the South; to greater competition and lower prices; to an international and genuinely market-based economy; to the notion of Northern countries living within their means.

Instead, attitudes have hardened. Debt, the result of investments made by government leaders and world bankers now gone, resulted in a staggering $104 billion flow of interest payments by aid recipients in 1997, double the aid bill that year from all bilateral and multilateral agencies combined.[10] Protection against the South continues in the form of tariffs and non-tariff trade barriers. International trade imbalances and price distortions continue to be nurtured by self-seeking Northern politicians, armed and advised by economists who demand self-righteously that Southern governments cease their economic experiments and start living within their means. Living within one's means is a convenient concept, but when the most basic means of life are withheld, when the price of cocoa continues to plummet, when exports are blocked from Northern markets, when jobs succumb to the capital intensity of Northern mass production and high technology, things continue to fall apart. As Yeats observed so powerfully, 'the centre cannot hold ... [and] anarchy is loosed upon the world.'

It doesn't have to be that way. Faustian man does not have to be dragged to death by his own machine, as Spengler predicted he would. Alternatives exist. There is still time. There is knowledge. There are enough committed people. Despite a cacophony of claims to the contrary, there is enough money. UNDP estimated in 1991 that if industrialized countries reduced their military expenditure by only three per cent, this would yield a savings of $25 billion a year. If developing countries froze their military expenditure, this would yield another $10 billion. The combined savings would be more than enough to eradicate malnutrition, to cover the cost of providing universal primary education, universal primary health care and safe drinking water by the year 2000.[11] Military expenditure did, in fact, decline. As a proportion of global GDP, it fell from 5.2 per cent in 1985 to 2.8 per cent in 1995.[12] But the savings were not devoted to health, education and clean drinking water. And the hoped-for changes, as elusive in 2000 as in 1991, had been postponed by the aid establishment to 2015 – a date far enough ahead to ensure that few of the nodding politicians and bureaucrats in Y2K would be held accountable for the inaction that will precede the next failure.

The test, as Robert Chambers says, is what people do. 'Social change flows from individual actions. By changing what they do, people move societies in new directions and themselves change.'[13] Because he started it all, perhaps it is best to allow Fritz Schumacher to have the final word. 'The starting point of all our considerations is poverty,' he wrote, 'or rather, a degree of poverty which means misery, and degrades and stultifies the human person.'[14]

> One way or another, everybody will have to take sides in this great conflict ... It is widely accepted that politics is too important a matter to be left to experts. Today, the main content of politics is economics, and the main content of economics is technology. If politics cannot be left to the experts, neither can economics and technology. The case for hope rests on the fact that ordinary people are often able to take a wider view, and a more 'humanistic' view, than is normally being taken by experts. The power of ordinary people, who today tend to feel utterly powerless, does not lie in starting new lines of action, but in placing their sympathy and support with minority groups which have already started.[15]

Notes

Where a source has been used more than once in the book, the full reference will be found in the bibliography.

Chapter I

1. UNDP: www.undp.org/info21/
2. Effusions about the miracle of the telegraph are taken from Standage, pp. 83, 103–4.
3. Rhodes, p. 70
4. Meadows *et al.*
5. Pearson, p. 149
6. *ibid.*, p. 116
7. *ibid.*, p. 72
8. Brandt, Willy, *et al.*, *North-South; A Programme for Survival*, MIT Press, Cambridge, Mass., 1980, p. 12
9. *ibid.*, p. 225
10. *ibid.*, p. 26
11. *ibid.*, p. 176
12. Brandt, Willy, *et al.*, *Common Crisis*, Pan Books, London, 1983, p. 107
13. World Bank, quoted in Tussie, Diana, *The Less Developed Countries and the World Trading System*, Pinter, London, 1987, p. 60
14. Brown and Eckholm, p. 97
15. Smillie (1986), p. 21
16. Schumacher (1974), pp. 165–6
17. Brandt (1980) *op. cit.*, p. 170
18. *ibid.*, p. 47
19. World Bank (1990), p. 127
20. *ibid.*, p. 128
21. World Bank (1997a), p. 134
22. World Bank (1990), p. 103
23. *WSJ*, 3 May 1990
24. Steven Holtzman, *Post-Conflict Reconstruction*, Washington, World Bank, Social Policy and Resettlement Division, Washington, 1997
25. Carnegie Commission on Preventing Deadly Conflict, *Preventing Deadly Conflict*, Carnegie Corporation, New York, 1997
26. These figures are taken from UNHCR's website, February 2000
27. Preliminary OECD/DAC estimates show an increase in ODA from $48.3 billion in 1997 (a figure adjusted from earlier totals), to $51.5 billion in 1998, an increase in real terms of 8.9 per cent. Source: *Development Information Update*, No. 1, July, 1999

28 See, for example, Foy, Colm and Helmich, Henny (eds) *Public Support for International Development*, OECD, Paris, 1996, and Smillie, Ian, and Helmich, Henny (eds.) *Public Attitudes and International Development Co-operation*, OECD, Paris, 1998
29 Cited in 'Foreign Direct Investment Flows to Low-Income Countries: A Review of the Evidence', *Briefing Paper 1997* (3), Overseas Development Institute, London
30 'Thailand's Tiger Economy Loses its Roar', Sandra Sugawara, *Washington Post*, 24 June 1997
31 World Bank (1997a), p. 49
32 *The Economist*, 25 July 1997
33 IMF, *World Economic Outlook*, Washington, 1997, quoted in *Globe and Mail*, Toronto, 17 September 1997
34 World Bank (1997a), p. 60
35 *International Herald Tribune*, 'World Bank Admits Failings in Indonesia's Collapse', 12 February 1999
36 OECD, 'Aid and Private Flows Fell in 1997', Press Release, Paris, 18 June 1998
37 OECD (1999), p. 49
38 *Wall Street Journal*, 'IMF Lowers its Forecast for Worldwide Growth', 22 December 1998
39 Mazower, Mark, *Dark Continent: Europe's Twentieth Century*, Knopf, New York, 1999, p. 82
40 Brown *et al.* (1999a), p. 120
41 Meadows, p. 17
42 *Globe and Mail*, Toronto, 'Afghan Site Promises Mother Lode of Copper', 28 November 1998
43 World Resources Institute (1996–7), p. 277. But then again, who knows? In 1998, geologists Colin Campbell and Jean Laherrere put total reserves at one trillion barrels in 'The End of Cheap Oil', *Scientific American*, March 1998
44 World Resources Institute (1990), pp. 84–5
45 World Resources Institute (1996), p. 238
46 Fuller, Buckminster, *Comprehensive Design Strategy*, World Resources Inventory, Phase II, University of Illinois, Carbondale, 1967, p. 48
47 OECD/DAC, *Shaping the 21st Century: The Contribution of Development Co-operation*, Paris, May, 1996
48 OECD (1999), p. 13

Chapter II

1 IMF, 'India: Recent Economic Developments', Washington, July 1989
2 Figures taken from World Bank (1990) and from Demery, L. and Walton, M., 'Are Poverty and Social Targets for the 21st Century Attainable?', paper prepared for a DAC/Development Centre Seminar on *Key Elements in Poverty Reduction*, Paris, 4–5 December 1997
3 Maxwell, Simon, *The Meaning and Measurement of Poverty*, ODI Poverty Briefing, Overseas Development Institute, London, 3 February 1999
4 Chambers (1983), pp. 48–9

[5] Hill, Polly, *Development Economics on Trial*, Cambridge University Press, Cambridge, 1986, p. 33
[6] Cox, Aidan and Healey, John, 'Promises to the Poor: The Record of European Development Agencies', *ODI Poverty Briefing*, ODI, London, 1 November 1998
[7] FAO, *The Dynamics of Rural Poverty*, Rome, 1986, p. 21
[8] Lipton, Michael, 'The Poor and the Poorest: Some Interim Findings,' *Discussion Paper No. 25*, World Bank, Washington, 1986, pp. 4–6
[9] Chambers (1988), p. 5
[10] Chambers (1988), pp. 8–9; See also Chambers (1983), Chapter 5
[11] Narayan, Deepa; Chambers, Robert; Shah, Meera; Petesch, Patti, *Global Synthesis; Consultations with the Poor*, Poverty Group, World Bank, 20 September 1999
[12] Figures on the HDI are taken from UNDP (1991), pp. 15–20
[13] UNDP (1998), p. 29
[14] Leonard, H.J., 'Environment and the Poor: Development Strategies for a Common Agenda', in Leonard *et al.*, p. 13
[15] *National Conditions Report*, cited in 'New revolution turns out victims', *Globe and Mail*, Toronto, 8 December 1998, p. A14
[16] Leslie, Joan, 'Women's Work and Child Nutrition in the Third World; Report for the Carnegie and Rockefeller Foundations', New York, 1987
[17] UNICEF (1999)
[18] World Bank (1997a), p. 220
[19] This common assumption may, however, be too glib. Jack Ives, Co-ordinator of the Mountain Ecology and Sustainable Development Project of the UN University, argues strongly that there is no clear link between land use in the mountains and changes in flows of sediment in the Ganges and Brahmaputra Rivers. See 'Floods in Bangladesh: Who is to Blame?', Jack Ives, *New Scientist*, April 1991
[20] Hobsbawm, p. 87
[21] Abidjan and Jakarta statistics taken from World Resources Institute (1996), pp. 4–7
[22] Cox and Healy, *op. cit.*, p. 4; OECD (1998), Table A45
[23] 'Latest World Bank Poverty Update Shows Urgent Need to Better Shield Poor in Crises', World Bank News release No. 99/2214/S, 2 June 1999
[24] World Bank (1997b), p. 43
[25] World Resources Institute (1996), p. 170
[26] Drucker, Peter, 'The Changed World Economy', *Foreign Affairs*, Spring, 1986
[27] Demery, L. and Walton, M., 'Are Poverty and Social Targets for the 21st Century Attainable?', Forum on Key Elements for Poverty Reduction Strategies, OECD, 4–5 December 1997, p. 22
[28] Quoted in Willoughby, p. 74

Chapter III

[1] Where development assistance is concerned, the most prominent theoreticians were P.N. Rosenstein-Rodan, 1961; 'International Aid for Underdeveloped Countries', in *Foreign Aid*, Bhagwati, J. and Eckaus, R.S. (eds) Penguin, 1970; W.W. Rostow, *Stages of Economic Growth*, 1960; *The Economics of Take-Off into Sustained*

    Growth, 1963; H. Chenery and A.M. Strout, 'Foreign Assistance and Economic Development', *American Economic Review*, September 1966
[2] Pearson, p. 52
[3] World Bank (1988), p. 2
[4] Brundtland, Gro Harlem, *et al.*, *Our Common Future: The World Commission on Environment and Development*, OUP, Oxford, 1987, p. 52
[5] World Resources Institute (1990), pp. 234–6
[6] World Resources Institute (1990), p. 236, and Brown *et al.*, (1990), p. 9
[7] World Bank (1990), p. 58
[8] World Bank (1990), p. 60
[9] Unless otherwise noted, statistical data on the Bangladesh transport sector are derived from Chapter 3 of *The Country Boats of Bangladesh,* by Jansen, Erik G., Dolman, A.J., Jerve, A.M., Rahman, N., University Press, Dhaka, 1989
[10] Sobhan, Rehman (ed.), *From Aid to Dependence: Development Options for Bangladesh*, University Press, Dhaka, 1990, p. 10; the figure is for the year 1982–3
[11] Jansen *et al.*, *op cit.*, p. 31
[12] Brown *et al.* (1990), p. 47
[13] Brown *et al.* (1997), p. 118
[14] 'The Dry Facts about Dams', *The Economist*, 20 November 1999
[15] Yudelman M., 'Sustainable and Equitable Development in Irrigated Environments', in Leonard *et al.*, p. 71
[16] Ryder, Grainne (ed.), Probe International, Toronto, 1990
[17] *ibid.*, p. 15
[18] 'China Admits a Third of its Dams near Collapse', *The Times*, 2 March 1999
[19] 'The Holes in China's Dam Project', *National Post*, Toronto, 18 June 1999
[20] Brown *et al.* (1998), p. 160
[21] Hertsgaard, Mark, *Earth Odyssey: Around the World in Search of Our Environmental Future*, Broadway Books, New York, 1999, p. 170
[22] *ibid.*, p. 181
[23] Riddell (1987), p. 235
[24] World Bank (1988), Annex I-A, para 41J
[25] *ibid.*, Annex II, p. 12
[26] Conway and Barbier, p. 77
[27] Riddell (1987), p. 237
[28] Farrington, J., 'Farm Power and Water Use in Sri Lanka', in Puttaswamaiah, K., (ed.), *Poverty and Rural Development; Planners, Peasants and Poverty*, Intermediate Technology Publications, London, 1990, p. 176
[29] Quoted in World Resources Institute (1990), p. 91
[30] *ibid.*
[31] *ibid.*, p. 92
[32] See 'Empowering Africa's Rural Poor', in Lewis, p. 89
[33] Quoted in Meadows (p. 29) and used as the title for a book on population by Lester Brown and Erik Eckholm
[34] Daly, Herman E., 'Sustainable Growth: An Impossibility Theorem', *Development*, Society for International Development, Rome, Vol. 3/4, 1990
[35] Quoted in Meadows *et al.* (1972)

Chapter IV

1. World Vision data are derived from World Vision International 1996 Annual Report, and from the 1996 Annual Reports of World Vision Australia, World Vision Canada and World Vision United States
2. Information derived from the annual reports of the NGOs cited
3. Annis, Sheldon, 'Can Small Scale Development be Large Scale Policy?' *World Development*, Vol. 15, Supplement, 1988
4. Salamon, Lester and Anheier, Helmut K., *The Emerging Sector Revisited; A Summary*, Center for Civil Society Studies, Johns Hopkins University, 1998, Table 1
5. de Tocqueville, Alexis, *Democracy in America*, originally published in 1835, republished by Doubleday Anchor, Garden City, 1969
6. Brown, David, 'Rhetoric or Reality? Assessing the Role of NGOs as Agencies of Grass Roots Development,' *Bulletin*, No. 28, University of Reading Agricultural Extension and Rural Development Department, Reading, 1990, p. 5
7. *ibid.*, p. 6
8. Korten, David, 'Third Generation Strategies; a Key to People-Centred Development,' *World Development*, Volume 15, Supplement, Oxford, 1987, and Korten, David, *Getting to the 21st Century*, Kumarian Press, West Hartford, 1990
9. Martin, Samuel A., *Financing Humanistic Service*, McClelland and Stewart, Toronto, 1975
10. BAIF was formerly an acronym for the Bharatiya Agro Industries Foundation
11. Satish, S. and Prem Kumar, N., 'Are NGOs More Cost-Effective than Government in Livestock Service Delivery?' in John Farrington *et al.* (eds) NGOs and the State in Asia, Routledge, London, 1993, pp. 169–71
12. Bahmueller, Charles, 'Civil Society is the Buzzword of the Hour', United States Information Agency Online Journal at http://civnet.org/index.html
13. Smillie and Henry
14. Salamon, Lester and Anheier, Helmut, 'The Third World's Third Sector in Comparative Perspective', in Lewis, pp. 70–1
15. Holloway, Richard, *Supporting Citizens' Initiatives*, Intermediate Technolgy Publications, London, 1998, p. 210
16. BRAC, *Annual Report 1997*, Dhaka
17. World Bank, *Financing Health Services in Developing Countries; An Agenda for Reform*, Washington, 1987, p. 42
18. AMREF material derived from Smillie, I., 'Strengthening NGOs: The Strangulation Technique', AMREF, Nairobi, 1988

Chapter V

1. Hawke, p. 8
2. Cited in Pacey, Arnold, *Technology in World Civilization*, MIT Press, Cambridge, Mass., 1990, p. 80
3. *ibid.*, pp. 114, 127
4. Mumford, Lewis, *Technics and Civilization*, Harcourt Brace, New York, 1939, pp. 14–15.

5. Harrison's dramatic story is told in Sobel, Dava, *Longitude*, Fourth Estate, London, 1996.
6. Boorstin, Daniel, *The Discoverers*, Vintage Books, New York, 1985, p. 201
7. Described in Best, p. 80
8. Pendle, George, *A History of Latin America*, Penguin, Harmondsworth, 1983, p. 41
9. Pacey, *op. cit.*, p. 147
10. The Kerala story is drawn from various ITDG reports, notably O'Riordan, Brian, 'Fisheries Development in South India; The Indo-Norwegian Fisheries Project', ITDG, 1990; from Kadappuram, John, 'Artificial Fishing Reef and Bait Technologies', in Gamser, M.S., Appleton, H., Carter, N. (eds) *Tinker Times Technical Change*, Intermediate Technology Publications, London, 1990, pp. 150–79; and from Kurien, John and Wilmann, Rolf, 'Economics of Artisanal and Mechanised Fisheries in Kerala; A Study on Costs and Earnings of Fishing Units', FAO, 1982. See also Le san, Alain, *A Livelihood from Fishing: Globalization and Sustainable Fisheries Policies*, Intermediate Technology Publications, London, 1998
11. Kerala fishing statistics through 1996 were provided by Satish Babu, South Indian Federation of Fishermen's Societies, February 1999
12. Cited in Gimpel, Jean, *The Mediaeval Machine; The Industrial Revolution of the Middle Ages*, Wildwood House, Aldershot, 1988
13. Noble, p. 53
14. Best, p. 71
15. The story is recounted in Best, p. 186.
16. Pacey, *op. cit.*, p. 120
17. Noble, p. 22
18. *ibid.*, p. 244
19. Quoted in Hawke, p. 77
20. See, for example, Drucker, Peter, *Adventures of a Bystander*, Harper & Row, New York, 1979, and Peters and Waterman, pp. 202–34
21. Peters and Waterman, p. 207
22. *ibid.*, p. 224
23. Study conducted by Alan Smithers and John Pratt, University of Surrey; reported in the Toronto *Globe and Mail*, April 8, 1991

Chapter VI

1. Quoted in Rybczynski, p. 41
2. Quoted in Willoughby, p. 64
3. Willoughby, p. 66
4. Quoted in Rybczynski, p. 37
5. Quoted in McRobie, pp. 24–5
6. *The Observer*, London, 29 August 1965, quoted in McRobie, pp. 25–31
7. McRobie, p. 36
8. 'Putting Partnership into Practice', ITDG, November 1989
9. National Film Board of Canada interview, 1977; quoted in Rybczynski, p. 165
10. Whiticombe, R. and Carr, M., *Appropriate Technology Institutions: A Review*, ITDG, London, 1982

11. Jade Mountain Website: www.jade-mtn.com
12. Woto, Teedzani, *Biogas Technology in Botswana; A Sociological Evaluation*, RIIC, Kanye, 1988
13. Willoughby reviews the literature and arguments critical of appropriate and intermediate technology, pp. 223–63
14. Chenery, H. *et al.*, *Redistribution With Growth*, Oxford University Press, Oxford, 1974
15. Schumacher (1974), p. 40
16. For example, Eckaus, R., *Appropriate Technologies for Developing Countries*, National Academy of Sciences, Washington, 1977; and 'Appropriate Technology: The Movement has Only a Few Clothes On,' in *Issues in Science and Technology*, Winter, 1987
17. 'Enabling', which I have used in the title for Part Three of this book, is one of those words that development workers have almost (but not quite) worn out. It was first used in the development context by H.H. The Aga Khan in a 1982 speech in Nairobi: 'Social Institutions in National Development: The Enabling Environment'.
18. Schumacher (1974), p. 223
19. *ibid.*, p. 133
20. Quoted in Best, p. 11
21. Best, p. 163
22. Peters and Waterman, p. 215
23. *The Economist*, 14 October 1989
24. Reich, p. 6
25. A good description of the 'Third Italy' can be found in Best, pp. 203–26
26. 'On the Tiles', *The Economist*, 2 January, 1999, p. 58
27. 'The Complications of Clustering', *The Economist*, 2 January, 1999
28. The Sialkot story is told in Nadvi, Khalid, 'International Competitiveness and Small Firm Clusters – Evidence from Pakistan', *Small Enterprise Development*, Vol. 9, No. 1, March 1998
29. The Ludhiana and Mexican examples are described in an MSc thesis by Michael Albu, 'Technological Learning and Innovation in Industrial Clusters in the South', Science Policy Research Unit, University of Sussex, September 1997
30. de Soto, Hernando, 'Structural Adjustment and the Informal Sector', in *Microenterprises in Developing Countries*, J. Levitsky (ed.) Intermediate Technology Publications, London
31. Hasan, Arif, 'The Informal City – Reinterpreting Karachi', *The News on Sunday*, Karachi, 8 November 1998
32. Gwitira, Joshua C., *Small Scale Technology for Agro-Industrial Development*, ITDG/Zimbabwe Ministry of Industry and Commerce, Harare, October, 1990
33. Discussion at Indigenous Business Development Centre, Harare, March, 1991
34. BBC World Service News, 15 February 1991
35. Smillie, Ian, *Relief and Development: The Struggle for Synergy*, Occasional Paper No. 33, Thomas J. Watson Jr. Institute for International Studies, Brown University, Providence, USA 1999, p. 17

36 Schumacher (1974), p. 138
37 Schumacher (1978), pp. 159–60

Chapter VII

1 Brown and Eckholm (1974), p. 137; International Rice Research Institute website: www.cgiar.org.irri, 1999
2 World Resources Institute (1996), pp. 238–9
3 See, for example, Lipton, Michael, with Longhurst, Richard, *Modern Varieties, International Agricultural Research, and the Poor*, CGIAR Study Paper No. 2, World Bank, Washington, 1985
4 World Resources Institute (1996), p. 235
5 An Interim Report on the State of the Forest Resources in the Developing Countries, FAO, Rome, 1988
6 Brown *et al.* (1998), p. 22
7 Leonard, p. 19
8 Brown *et al.* (1998), p. 30
9 World Resources Institute (1990), p. 107
10 Worldwide Fund for Nature, *Living Planet Report*, Gland, 1998, p. 6
11 Brown *et al.* (1998), p. 25
12 World Resources Institute (1996), p. 170
13 'Commodity Price Data Pink Sheet – November 1999', World Bank, Washington
14 Conway and Barbier, p. 76
15 Leonard, p. 38–9. See also 'ICRAF Online', website of the Kenya-based International Council for Research in Agricultural Forestry: www.cgiar.org/icraf, 1999
16 Details gathered during the author's personal visit
17 Conroy, Czech and Litvinoff, Miles, *The Greening of Aid: Sustainable Livelihoods in Practice*, Earthscan, London, 1988, p. 81
18 World Resources Institute (1990), p. 111
19 Brown *et al.* (1991), p. 181
20 Smillie, p. 138–41
21 World Resources Institute 1990, p. 106
22 Browder, John O., 'Development Alternatives for Tropical Rainforests', in Leonard *et al.* p. 117
23 Brown *et al.* (1991), p. 80
24 Details on the Wokabout Somil are taken from Unwin, Paul, *The Wokabout Somil: A New Rural Industry for Papua New Guinea*, Appropriate Technology, London, December, 1990
25 BRAC, Annual Report 1997, Dhaka, p. 21
26 The description of this project has been taken from discussions with ITDG staff and from ITDG documents, including *Kenya Livestock Programme; A Review of the First Three Years of Operation, 1986–9*
27 Description of the Turkana Rehabilitation Programme is derived from discussions with ITDG, from a 1990 review by Cathy Watson and Arupe Lobuin entitled, 'Lokitaung Pastoral Development Project: Lakeshore Study', and from a book by Adrian Cullis and Arnold Pacey, *A Development Dialogue:*

Rainwater Harvesting in Turkana, Intermediate Technology Publications, London, 1992
28 Broche-Due, V., *Women at the Backstage of Development*, FAO Consultant's Report GCP/KEN/048/NOD, 1983; cited in Cullis and Pacey, 1992, *op. cit.*
29 Watson, Cathy and Ndung'u, Beth, 'Rainwater Harvesting in Turkana: An Evaluation of Impact and Sustainability', ITDG, Nairobi, November 1997
30 R.S. Hogg reviewed the pre-1980 projects and the post-*Lopiar* efforts over a ten-year period. See, for example, *Destitution and Development: A Strategy for Turkana*, (Dept of Anthropology, University of Manchester, 1982); *Water Harvesting in Semi-Arid Kenya: Opportunities and Constraints,* (Oxfam, 1986); *Building Pastoral Institutions: A Strategy for Turkana District,* (Oxfam, 1987); 'Water Harvesting and Agricultural Production in Semi-Arid Kenya', *Development and Change*, Vol. 19, 1988
31 Conway and Barbier, p. 37

Chapter VIII

1 Material on drying technologies has been derived from Barrie Axtell *et al.*, *Try Drying It*, Intermediate Technology Publications, 1991; UNIFEM Food Cycle Technology Source Books, New York, 1988, and ITDG files
2 Greeley, p. 11
3 Dolorier's story is told in Viani, Bruno, *Tray Drier Dissemination in Peru*, ITDG, 1996
4 Hagelberg, Gerry, 'The Structure of World Production and Consumption', in Kaplinsky *et al.* (1989), p. 49. 'Sugar' here means sugar produced by the OPS or VP techniques as opposed to the traditional coarse dark sugar known as jaggery
5 Kaplinsky (1990), p. 128.
6 The proceedings of a 1987 ITDG-sponsored conference on sugar provide extensive background detail as well as technical and economic analysis of the OPS technique. The publication (Kaplinsky *et al.*, 1989) is an invaluable source of background information for anyone interested in pursuing the subject further
7 Sunga, Ishmael and Whitby, Garry, *Decentralised Edible Oil Milling in Zimbabwe: An Evaluation Report of the Tinytech Oil Mill project*, ITDG, Zimbabwe, May 1995
8 The Bielenberg Ram Press story is drawn from the proceedings of and papers presented at a conference on the Bielenberg Ram Press and Small-Scale Oil Processing, held in Nairobi by ATI in September, 1990
9 Hyman, Eric L., *Preliminary Socio-Economic Impact Study of the Ram Press in the Arusha Region of Tanzania*, ATI, Washington, 1990
10 IDRC, 'Small-scale Rural Oilseed Processing in Africa', IDRC Website, www.idrc.ca/nayudamma/oilseed, June 1998. Ironically, while IDRC's website records ATI's success, there is no record there or among IDRC headquarters staff of its own experience
11 Comments made by Carlos Zulberti, IDRC Consultant, VOPS-Kenya Project, Report on an ATI Oil Press Conference, Nairobi, September, 1990

12 Muthaka, James K., 'Vegetable Oil/Protein System Programme; The Kenyan Experience', Paper presented at Bielenberg Ram Press Conference, Nairobi, September, 1990

Chapter IX

1 *Life*, 1932, quoted in *Bartlett's Familiar Quotations*, Little Brown and Company, Boston, 1992, p. 555
2 Inger, David, 'Overview of Woodstove Development in Botswana', in Caceres, R. (ed.), *Stoves for People*, Intermediate Technology Publications, London, 1989
3 Barnett, Andrew, the Diffusion of Energy Technology in the Rural Areas of Developing Countries: A Synthesis of Recent Experience', *World Development*, Vol. 18, No. 4, 1990, p. 540
4 'Commodity Price Data Pink Sheet – November 1999', World Bank, Washington
5 DFID, *Eliminating World Poverty – A Challenge for the 21st Century; Summary, White Paper on International Development*, HMSO, London, 1997, p. 11
6 For example, Joseph, S., 'An Appraisal of the Impact of Improved Wood Stove Programmes', in Caceres (ed.) *op cit*. The Caceres book contains several case studies which bear out this finding
7 'Indoor Air Pollution in Developing Countries and Acute Respiratory Infection in Children,' Pandey, M.R. *et al.*; the article, which appeared in the 25 February 1989 *Lancet*, was reprinted in edited form in *Boiling Point*, No. 18, ITDG, Rugby, April, 1989
8 Crewe, Emma and Harrison, Elizabeth, *Whose Development? An Ethnology of Aid*, Zed Books, London, 1998, p. 100
9 'How to Make the Kenyan Ceramic Jiko', Kenya Renewable Energy Project, Ministry of Energy/EDI, 1983; other background on the KCJ is taken from ITDG reports and from 'The Introduction of the Kenya Jiko Stove; A Kengo Experience', by Opole, Monica, in Carr (1988)
10 The Sri Lankan stoves material is derived from visits to Negombo stove makers and retailers, from discussions with CEB officials, and from various internal ITDG studies and reports; also from 'The Sarvodaya Stoves Project: A Critical Review of Developments 1979–82', by M. Howes *et al.*; 'A History of the Sarvodaya Stoves Project 1979–82', by Bill Stewart, both Intermediate Technology Publications, London 1983; and Crewe, Emma, Young, Peter, and Sundar, Shyam, *A History of Sri Lanka Stoves Programmes* (draft), ITDG, Rugby, 1998
11 Amarasekera, R.M., 'Status of Improved Woodstove Dissemination in Sri Lanka', in Caceres (ed.) *op cit.*, p. 118
12 *Sri Lanka Bilateral Woodburning Stoves Project; Internal Evaluation*, ITDG, 1989
13 *ibid.*, p. 25
14 Calculations based on a cost analysis in Crewe *et al.*, *op cit.*, p. 27
15 Von Weizäcker *et al.*, p. 47
16 Advani, S.B., *Solar Photovoltaic for Sustainable Development*, Crompton Greaves Ltd, Nashik, India, 1999
17 Brown *et al.* (1998), p. 179
18 Sinha, p. 20

19 World Bank (1994), p. 30
20 'Argentine city crippled by outage', Toronto *Globe and Mail*, 23 February, 1999
21 Windmills in Argentina are discussed at length in Hurst, C., 'Establishing New Markets for Mature Energy Equipment in Developing Countries', *World Development*, Vol. 18, No. 4, 1990
22 'US Aims to Have 5% of Electricity From Wind by 2020', *New York Times*, 20 June 1999
23 Foley, p. 36
24 These figures are provided by Flavin, p. 50. There is ambiguity about the numbers, however. For example Hurst, *op. cit.*, uses the much larger figure of 90 000 small plants by 1980, with a capacity of 10 000 megawatts
25 Flavin, p. 47
26 Hagler, Baily and Co., 'Decentralized Hydropower in AID's Development Assistance Program', quoted in Flavin, p. 49
27 Material derived from ITDG reports and studies, from Hurst and Flavin, *op. cit.*, and from Hislop, D., *The Micro Hydro Programme in Nepal – A Case Study*, Biomass Energy Services and Technology, London, 1987
28 Hislop, Drummond, *Upgrading Micro Hydro in Sri Lanka*, Intermediate Technology Publications, London, 1985
29 Mini Hydro Power Group, 'Reducing Global Warming', FAKT, Furtwangen, Germany (undated)
30 *ibid.*
31 The Sri Lanka story is derived from discussions with ITDG staff and others in Britain and Sri Lanka, from ITDG studies and reports, from Holland, R. and Hislop, D., *Upgrading Micro Hydro in Sri Lanka*, Intermediate Technology Publications, London, 1986; and from *Mini Hydro: A Sound Investment*, Mini Hydro Power Group, FAKT, Furtwangen, Germany (undated)
32 Barnett, Andrew, 'The Diffusion of Energy Technology in the Rural Areas of Developing Countries: A Synthesis of Recent Experience', *World Development*, Vol. 18, No. 4, 1990

Chapter X

1 Agevi, E. *Paradoxes in the Utilization of Local Building Resources for Low Cost Housing Schemes in Nairobi*, Nairobi, 1988
2 Government of Malawi National Statistics Office, cited in *Balaka Lime Pilot Project Evaluation*, READI, Lilongwe, 1991
3 Details of this project have been derived from *Balaka Lime Project Evaluation*, READI, Lilongwe, 1991; *Review of Chenkumbi Limeworks Limited*, PPI Consultants Limited, Lilongwe, 1994, and personal communication with Garry Whitby, 1999
4 'Minister Pillane in Financial Scam', *Daily Times*, Blantyre, Malawi, 7 February, 1997
5 Rondinelli, Dennis, *Development Projects as Policy Experiments*, Routledge, 1993, pp. 106–7
6 Sinha, p. 128

7 *ibid.*, p. 75
8 Sinha, p. 134
9 Kaplinsky (1990), p. 196
10 Robert L. Day, *Pozzolans for Use in Low-Cost Housing*, IDRC, Ottawa, 1990
11 This section on brickmaking in Botswana is based on work done by Raphael Kaplinsky (1990), pp. 74–103, and on *ATI's Brickmaking Projects in Three African Countries: Mali, Botswana, Tanzania*, Carlos Lola, ATI, Washington, 1987
12 Kaplinsky (1990), p. 83
13 Details on Zimbabwe brick production are taken from Mason, Kelvin, *Small-Scale Brick Firing: The Experience of ITDG in Zimbabwe*, ITDG, Peru, 1997, and Schilderman, Theo, *Sustainable Materials Production: A Question of Energy?*, ITDG, Rugby, 1998
14 I am grateful to many people for the background to this story. They include Solomon Mwangi, Nick Evans, Nick Hall and Shelter Works in Kenya, Sosthenes Buatsi at the Technology Consultancy Centre in Ghana, and John Parry in Cradley Heath. Publications consulted include Evans, Barrie, *Understanding Natural Fibre Concrete*, Intermediate Technology Publications, London, 1987; and Gram, H.E. *et al.*, Fibre Concrete Roofing, SKAT, St Gallen, 1987
15 Interview with John Parry, Cradley Heath, February, 1991
16 Adapted from *A Ten Point Prescription for a Good Agent for Parry Equipment*, J.P.M. Parry & Associates Ltd., Cradley Heath
17 Coughlin, Peter, 'Steel Versus Tile Roofing: What's Appropriate for Kenya?' Industrial Research Project, University of Nairobi, August 1985

Chapter XI

1 Described by Ohiorhenuan, John F.E. in *The Industrialization of the Very Late Starters: Historical Experience, Prospects and Strategic Options for Nigeria*, IDS Discussion Paper No. 273, Brighton, 1990
2 Fieldhouse, D.K., *Black Africa 1945–1980; Economic Decolonization and Arrested Development*, Allen & Unwin, London, 1986, p. 232
3 Smillie, p. 8
4 Karmiloff, Igor, 'Zambia', in Riddell (1990), p. 313
5 Stevens, Christopher, 'Nigeria', in Riddell (1990), p. 273
6 Riddell (1990), p. 374; correspondence with ITDG Harare, April 1999
7 OECD (1993) and (1999)
8 Stevens, *op. cit.*, p. 256
9 World Bank (1999)
10 Ndlela, Daniel B., 'Macro-policies for Appropriate Technology in Zimbabwe Industries', in Stewart, p. 179
11 O'Regan, Fred, *et al.*, 'Malawi: Informal Sector Assessment', prepared for Labat-Anderson Inc., and for AID/AFR/MDI, 1989
12 I met Frank Soko in Malawi in 1990 but could not find him on a subsequent visit in 1996. In 1999 I was told that he had died from AIDS
13 The sections of this chapter dealing with Ghana, the Technology Consultancy Centre, GRATIS and SIS Engineering are drawn from Smillie (1986);

notes, internal documents and extensive discussions in Ghana with the principals of these institutions in February, 1991; from an interview with John Powell in 1999; and reports supplied by CIDA and TCC in 1999
14. Adjorlolo, B.T., *Growing From a Small Beginning; A Handbook on How to Start and Manage a Small-Scale Industry*, University Press, Kumasi, 1985
15. Interview with John Powell, Oxford, January 1999
16. For a full description of the CIDA and USAID debacles, see the chapter entitled 'Partners in Development' in Smillie, pp. 72–82
17. GRATIS, 'Impact Assessment Report for the Year Ending December 31, 1998', Tema, February 1999
18. Hobsbawm, p. 173

Chapter XII

1. FACET BV, 'Evaluation of GRATIS', August, 1996, Utrecht, p. 4
2. Goss-Gilroy Inc., 'Mid-Term Evaluation of GRATIS (Phase II)', Ottawa, December 1997, p. 1
3. Ward, Barbara, and Dubos, René, *Only One Earth*, Penguin, Harmondsworth, 1972
4. IUCN, *Achievements 1978–81* (Report of the Director General), IUCN, Gland, 1981
5. World Commission on Environment and Development, *Our Common Future*, OUP, Oxford, 1987, p. 45
6. ul Huq, Mahbub, 'Employment in the 1970s; A New Perspective', *International Development Review*, 1971, Vol. 4, pp. 9–13
7. UNDP, *Human Development Report 1994*, OUP, New York, 1994, p. 4
8. Paddock, William and Elizabeth, *We Don't Know How: An Independent Audit of What They Call Success in Foreign Assistance*, Iowa State University Press, Ames, 1973
9. Adapted from Canadian Public Health Association, *Helping Children Beat the Odds: Lessons Learned from Canada's International Immunization Program – Phase 2*, Ottawa, 1997, p. 52
10. For a good discussion of the experimental nature of development assistance, risk management and learning, see Rondinelli, Dennis, *Development Projects as Policy Experiments*, Routledge, London, 1993
11. *The Strategic Compact: Renewing the Bank's Effectiveness to Fight Poverty*, World Bank, February, 1997
12. Riddell, Roger, *Aid in the 21st Century*, Office of Development Studies, UNDP, New York, 1996, p. 31
13. Handy, Charles, *The Hungry Spirit*, Hutchinson, London, 1997, p. 8
14. OECD (1996)
15. Robert Chambers frequently writes on this theme. A good discussion of the subject can be found in *Challenging the Professionals; Frontiers for Rural Development*, Intermediate Technology Publications, London, 1993
16. *ibid.*
17. Riddell, *op. cit.*, pp. 31–2

18. Daly, Herman, *Beyond Growth: The Economics of Sustainable Development*, Beacon Press, Boston, 1996, p. 27

Chapter XIII

1. Fernando, Priyanthi, 'Gender and Transport', in Everts, Saskia, *Gender and Technology*, ToolConsult and Zed Books, Amsterdam and London, 1998, p. 131–2
2. Andam, Aba. A., 'Remedial Strategies to Overcome Gender Stereotyping in Science Education in Ghana', in *Development*, SID, Rome, 1990
3. UNDP (1998), p. 136
4. Tendler, J., quoted in 'Whatever Happened to Poverty Alleviation', in Levitsky (1989)
5. McKee, Katharine, 'Microlevel Strategies for Supporting Livelihoods, Employment and Income Generation of Poor Women in the Third World: The Challenge of Significance', *World Development*, Vol. 17, No. 7, 1989
6. The CARE case is based on personal study by the author during a visit to the project in August 1990
7. Harper *et al.*, p. 64 Intermediate Technology
8. *ibid.*, p. 77
9. McKee, *op. cit.*, p. 1001
10. Brian Rowe, personal communication
11. Figure as of 1999; source: BRAC Annual Report, Dhaka, 1999, p. 10
12. Carr (1984), pp. 13–14
13. Chen, Marty, 'A Sectoral Approach to Women's Work: Lessons from India', *World Development*, Vol. 17, No. 7, 1989
14. *ibid.*, p. 60
15. Marsden, Keith, 'African Entrepreneurs; Pioneers of Development', *Discussion Paper No. 9*, International Finance Corporation, World Bank, Washington, 1990, p. 49
16. Kaplinsky (1990), p. 182
17. *ibid.*, p. 178
18. *Manual Silk Reeling in South India: A Preliminary Study of Technology Transfer*, Economic Development Associates and ITDG, Intermediate Technology Publications, London, 1990
19. Statistics from 1997 BRAC Annual Report, Dhaka, p. 23
20. BRAC RDP Report, June 1990, other BRAC reports; author's visit to BRAC and project sites. Devastating floods in 1998, and the arrival of low-priced Chinese yarn, damaged the project somewhat, but the number of silkworm rearers stabilized at 15 000 in 1999
21. Carr (1984) p. 6
22. McKee, *op cit.*
23. Creevey, Lucy, *Changing Women's Lives and Work*, Intermediate Technology Publications, 1996, p. 144
24. Ahmad M. and Jenkins, A., 'Traditional Paddy Husking – An Appropriate Technology Under pressure', *Appropriate Technology*, Vol. 7, No. 2, 1980; Greeley (1991)

25. Carr, Marilyn, Chen, Martha, Jhabvala, Renana (eds) *Speaking Out: Women's Economic Empowerment in South Asia*, Intermediate Technology Publications, London, 1996, p. 213

Chapter XIV

1. Jolly, R. *et al.* (eds) *Third World Employment*, Penguin, 1973, p. 9
2. Leonard *et al.*, p. 13
3. ILO, *World Employment Report 1998–99*, www.ilo.org
4. World Bank (1999)
5. Harrison, Paul, *The Third World Tomorrow*, Pelican, 1980, pp. 136–7
6. ILO, *op. cit.*
7. Cisneros, Susana Pinilla, 'Supporting Women in the Informal Sector: A Peruvian Experience', in *Women in Micro- and Small-Scale Enterprise*, Louise Dignard and José Havet (eds), Westview Press, Boulder, 1995, p. 164
8. *Conditions for Social Progress; Humane Markets for Humane Societies*, Ministry of Foreign Affairs, Copenhagen, 1997, p. 14
9. 'Bad Times in Brazil', *Globe and Mail*, Toronto, 27 July 1999
10. Statistics South Africa, *Unemployment and Employment in South Africa*, quoted in Hossack, Colin, *Informal sector employment 50% bigger than previously thought*, WOZA Internet (Pty) Ltd, 8 October 1998
11. Karachi details taken from 'The Informal City – Reinterpreting Karachi' by Arif Hasan, *The News on Sunday*, Karachi, 8 November 1998
12. Nowak, Maria, 'The Role of Microenterprises in Rural Industrialization in Africa', in Levitsky, p. 70
13. de Soto, Hernando, 'Structural Adjustment and the Informal Sector', in Levitsky, p. 3
14. Tilak, J.B.G., 'Education and its Relation to Economic Growth, Poverty and Income Distribution: Past Evidence and Further Analysis', World Bank, Washington, 1989. See also World Bank (1999), pp. 18–21
15. *ibid.*, pp. 24, 25
16. UNICEF (1999), p. 7
17. OECD (1999), Table 19: Aid by Major Purposes, 1996
18. World Bank (1999), Tables 6 and 17
19. UNICEF (1999), p. 108
20. World Bank (1999), p. 45
21. Dawson, J., *Small-Scale Industry Development in Ghana: A Case Study of Kumasi*, for Presentation to ESCOR, ODA, London, 1988
22. Mokyr, p. 264
23. King, Kenneth, 'Education for Employment and Self-Employment Interventions in Developing Countries: Past Experience and Present Prognosis', in *Defusing the Time Bomb?* International Foundation for Education with Production, Gaberone, 1990, p. 105
24. Reich, p. 229
25. Reno, William, *Corruption and State Politics in Sierra Leone*, Cambridge University Press, Cambridge, 1995, p. 138

26 Quoted in Reno, *op. cit.*, p. 146
27 de Soto, p. 133–6.
28 de Soto, in Levitsky, p. 5
29 Harper *et al.*, p. 4
30 Carr (1988), p. 8

Chapter XV

1 World Bank, 'Making Adjustment Work for the Poor', Washington, 1990, p. 5
2 'BMW to invest £3.3bn in Rover over six years', *The Independent*, 24 June 1999. The government offer lapsed in March 2000 when BMW jettisoned Rover and sold off Land Rover to Ford.
3 'Corporate Welfare', *Time Magazine*, 9 November 1998
4 Kristoff, Nicholas D., 'At This Rate, We'll Be Global in Another Hundred Years', *New York Time*s, 23 May 1999
5 *ibid.*
6 Soros, George, *The Crisis of Global Capitalism*, Public Affairs, New York, 1998, p. xxvi
7 *ibid.*, p. 204
8 The author is grateful to Siapha Kamara for information gleaned during a 1999 comparative shopping expedition to various drinking establishments in Wa, a town in northern Ghana
9 See Wells, L.T., 'Economic Man and Engineering Man: A Choice of Technology in a Low Wage Country', *Public Policy*, Vol. 21, No. 3, 1973; James, J., 'Appropriate Technology in Publicly Owned Industries in Kenya and Tanzania' in Stewart (1990) and Kaplinsky (1990)
10 James, *op. cit.*, p. 348
11 Theobald, Robin, *Corruption, Development and Underdevelopment*, Macmillan, 1990, p. 157
12 de Soto, p. 134
13 Source: European Commission, quoted in World Bank 1999, p. 2
14 White, J., 'External Development Finance and the Choice of Technology' in James, J. and Watanabe, S., *Technology, Institutions and Government Policies*, Macmillan, London, 1985
15 CIDA also supports CARE's Rural Maintenance Project, described in Chapter XII. This project involves more than 30 000 women in rural road maintenance
16 World Bank, 'Appropriate Technology in World Bank Activities', Washington, 1976, pp. vi and 8
17 World Bank (1999), p. 2
18 Standage, p. 83
19 UNDP (1999), p. 59
20 Postman, p. 93
21 Riddell (1990), p. 3
22 Hancock (1989), p. 171
23 Joint BIS-IMF-OECD-World Bank statistics on external debt; www.oecd.org/dac/debt

24 Young, Roger, *Canadian Development Assistance to Tanzania*, North South Institute, Ottawa, 1983, p. 77
25 Author's observation and discussions with CIDA and GRATIS staff at Tamale ITTU, February, 1991
26 Quoted in Noble, p. viii
27 Sachs, Jeffrey, 'Helping the World's Poorest', *The Economist*, 14 August 1999
28 Stewart, pp. 14–15
29 Bautista, Romeo M., 'Macro Policies and Technical Choice in the Philippines', in Stewart, p. 126
30 Government of Costa Rica MIDEPLAN, quoted in Doryan-Garron, E., 'Macroeconomic Policy, Technological Change and Rural Development: The Costa Rica Case', in Stewart, pp. 225, 229
31 Braun, Ernst, *Wayward Technology*, Frances Pinter, London, 1984, p. 100
32 Mokyr, pp. 173, 198–9
33 The prices were influenced in part by a deposit required on bottles, which were in short supply. The aluminium container of the import also had a certain resale scrap value. In fact, however, the price differential remained enormous even if the drink was consumed in a bar and the containers played no part in the transaction
34 *Botswana Science & Technology: A Policy and Implementation Framework*, Maendeleo, Botswana Technology Centre, Gaberone, 1990

Chapter XVI

1 UNICEF, 1998, p. 11
2 In fact some agencies *had* experimented with the mechanization of country boats. In 1973, the author was involved in a CARE project that experimented with Thai long-shaft engines in two large boats. Both sank during trials, as did the project. However, given the cost of imported engines and fuel, and given the new pollution from motorized rivercraft, the success of others and the end of the sailing era is perhaps not something to celebrate
3 South Commission Report, *The Challenge to the South*, OUP, New York, 1990
4 Chambers (1983), p. 190
5 Verne, Jules, *Paris in the Twentieth Century*, Ballantine Books, New York, 1997
6 Rhodes, p. 52
7 *ibid.*, pp. 130–1
8 Postman, p. 5
9 Steinbeck, John, *The Log From the Sea of Cortez*, 1951, quoted in Klitgaard, Robert, *Tropical Gangsters*, Basic Books, New York, 1990, p. 13
10 OECD (1999), Table A1
11 UNDP (1991), p. 10; similar estimates were made by the 1991 Stockholm Initiative on Global Security and Governance
12 World Bank (1999), p. 223
13 Chambers (1983), p. 217
14 Schumacher (1974), p. 140
15 *ibid.*, pp. 131–2

Bibliography

Additional references, for sources used in only one chapter, are found in the notes at the end of each chapter.

Best, Michael H., *The New Competition*, Polity Press, Cambridge, 1990
Brown, Lester and Ekholm, Erik, *By Bread Alone*, Praeger, New York, 1974
Brown, Lester, Gardner, Gary and Halweil, Brian, *Beyond Malthus*, W.W. Norton & Co., New York, 1999a
Brown, Lester, *et al.*, *State of the World, 1989, 1990, 1991, 1997, 1998, 1999*, Norton & Co., New York, 1989, 1990, 1991, 1997, 1998, 1999b
Carr, Marilyn, *Blacksmith, Baker, Roofing-sheet Maker: Employment for Rural Women in Developing Countries*, Intermediate Technology Publications, London, 1984
Carr, Marilyn (ed.) 1988, *Sustainable Industrial Development; Seven Case Studies,* Intermediate Technology Publications, London, 1988
Chambers, Robert, *Rural Development; Putting the Last First*, Longman, Harlow, UK, 1983
Chambers, Robert, *Poverty in India: Concepts, Research and Reality*, IDS Discussion Paper, No. 241, Sussex, 1988
Conway, Gordon R. and Barbier, Edward, *After the Green Revolution: Sustainable Agriculture for Development*, Earthscan, London, 1990
de Soto, Hernando, *The Other Path*, Harper & Row, New York, 1989
Flavin, C., *Electricity for a Developing World*, Worldwatch Paper No. 70, Washington, 1986
Foley, Gerald, *Electricity for Rural People*, Panos, London, 1989
Greenley, Martin, *Postharvest Technologies (!) Implications for Food Policy Analysis*, Economic Development Institute of the World Bank, Washington, 1991
Hancock, Graham, *Lords of Poverty*, Atlantic Monthly Press, New York, 1989
Harper, Malcolm, Esipisa, Ezekiel, Mohanty, A.K. and Rao, D.S.K., *The New Middlewoman. Profitable Banking through Ch-Lending Groups*, Intermediate Technology Publications, London, 1998, p. 64
Hawke, David Freeman, *Nuts and Bolts of the Past*, Harper & Row, New York, 1988
Hobsbawm, E.J., *Industry and Empire*, Penguin, London, 1969
Kaplinsky, Raphael, *The Economies of Small; Appropriate Technology in a Changing World*, Intermediate Technology Publications, London, 1990
Kaplinsky, Raphael *et al.*, *Cane Sugar; The Small-Scale Processing Option*, Intermediate Technology Publications, London, 1989
Leonard, H. Jeffrey *et al.*, *Environment and the Poor: Development Strategies for a Common Agenda*, Overseas Development Council, Washington, 1989
Levitsky, Jacob (ed.), *Microenterprises in Developing Countries*, Intermediate Technology Publications, London, 1989
Lewis, David (ed.), *International Perspectives on Voluntary Action*, Earthscan, London, 1999

McRobie, George, *Small is Possible*, Jonathan Cape, London, 1981
Meadows, Donella H. et al., *The Limits to Growth*, Earth Island, London, 1972
Mokyr, Joel, *The Lever of Riches; Technological Creativity and Economic Progress*, OUP, New York, 1990
Noble, David F., *America By Design; Science, Technology and the Rise of Corporate Capitalism*, OUP, New York, 1979
OECD, *Development Co-operation 1988, 1992, 1995, 1997, 1998*, DAC, Paris, 1989, 1993, 1996, 1998, 1999
Pacey, Arnold, with Cullis, Adrian, *Rainwater Harvesting*, Intermediate Technology Publications, London, 1986
Pearson, Lester B. et al., *Partners in Development*, Praeger, New York, 1969
Peters, T.J. and Waterman, R.H., *In Search of Excellence: Lessons From America's Best-Run Companies*, Warner Books, New York, 1982
Postman, Neil, *Building a Bridge to the Eighteenth Century: How the Past can Improve our Future*, Knopf, New York, 1999
Puttaswamaiah, K. (ed.), *Poverty and Rural Development; Planners, Peasants and Poverty*, Intermediate Technology Publications, London, 1990
Reich, Robert B., *The Work of Nations; Preparing Ourselves for 21st Century Capitalism*, Alfred Knopf, New York, 1991
Rhodes, Richard, *Visions of Technology*, Simon & Schuster, New York, 1999
Riddell, R.C., *Foreign Aid Reconsidered*, James Currey, London, 1987
Riddell, R.C., *Manufacturing Africa*, James Curry, London, 1990
Rybczynski, W., *Paper Tigers*, Prism, Chalmington, Dorchester, 1980
Ryder, Grainne, *Damming the Three Gorges*, Probe International, Toronto, 1990
Schumacher, E.F., *Small is Beautiful*, Abacus, London, 1974
Schumacher, E.F., *A Guide for the Perplexed*, Abacus, London, 1978
Sinha, Sanjay, *Mini-Cement; A Review of the Indian Experience*, Intermediate Technology Publications, London, 1990
Smillie, Ian, *No Condition Permanent; Pump-Priming Ghana's Industrial Revolution*, Intermediate Technology Publications, London, 1986
Smillie, Ian and Helmich, Henny (eds), *Stakeholders: Government–NGO Partnerships for International Development*, Earthscan, London, 1999
Standage, Tom, *The Victorian Internet*, Walker, New York, 1998
Stewart, Frances, Thomas, Henk and de Wilde, Ton, *The Other Policy*, Intermediate Technology Publications, London, 1990
UNDP, *Human Development Report 1990, 1991, 1998, 1999*, OUP, New York, 1990, 1991, 1998, 1999
UNICEF *The State of the World's Children 1989, 1990, 1991, 1999*, UNICEF, New York, 1989, 1990, 1991, 1999
Von Weizsäcker, Ernst, Lovins, A.B. and Lovins, L.H., *Factor Four: Doubling Wealth, Halving Resource Use*, Earthscan, London, 1998
Willoughby, Kelvin W., *Technology Choice; A Critique of the Appropriate Technology Movement*, Intermediate Technology Publications, London, 1990
Wood, Barbara, *Alias Papa: A Life of Fritz Schumacher*, Jonathan Cape, London, 1984
World Bank, *World Development Report 1980*, World Bank, Washington, 1980

World Bank, *Accelerated Development in Sub-Saharan Africa*, World Bank, Washington, 1981

World Bank, 'Report of the Task Force on Poverty Alleviation', Vols I and II, mimeo, Washington, 1988

World Bank 1989a, *Sub-Saharan Africa; From Crisis to Sustainable Growth*, World Bank, Washington, 1989

World Bank 1989b, *World Development Report 1989*, World Bank, Washington, 1989

World Bank, *World Development Report 1990: Poverty*, World Bank, Washington, 1990

World Bank, *World Development Report 1994: Infrastructure for Development*, World Bank, Washington, 1994

World Bank, *World Development Report 1995: Workers in an Integrating World*, World Bank, Washington, 1995

World Bank, *World Development Report 1997: The State in a Changing World*, World Bank, Washington, 1997a

World Bank, *Global Development Finance*, Vol. 1, World Bank, Washington, 1997b

World Bank, *World Development Report 1998/99: Knowledge for Development*, World Bank, Washington, 1999

World Resources Institute, *World Resources 1990–91*, OUP, New York, 1990

World Resources Institute, *World Resources 1996–97*, OUP, New York, 1996

Index

absolute poverty 24, 25, 32, 106, 215
accountability 62
ActionAid 164, 165–6
Adjorlolo brothers 179–81
adjustment 9–10, 14, 226–8
Africa
    industrialization failures 171–4
    poverty 27
African Medical and Research Foundation (AMREF) 61–2
Aga Khan Rural Support Programme 55
Agricultural Development Bank of Nepal (ADBN) 149
agriculture
    animal husbandry 113–14
    green revolution 104–6
    research 233
    sustainable 107–9
agroforestry 109–13
aid
    decline 16–17
    limitations 259
    private capital 17–19
    targets 7, 9, 13
    tied 236–7
aid fatigue 248
AIDS 19, 177
Akosombo Dam 40–1
alternative technology 93, 138–9
aluminium casting 176–7, 183–4
Amarasekera, R.M. 142–5
Annis, Sheldon 49
Applied Energy Services (AES) 111
apprenticeships 219–20
appropriate technology (AT) 90–8, 127, 249–50
Appropriate Technology Development Association (ATDA) 125, 157
Appropriate Technology International (ATI) 92, 97, 131–3
APPROTECH 93
Aral Sea 41
Argentina, windmills 147
Arkwright, Richard 81
arms race 71–2
arms trade 15, 261
Asia, poverty 27

assembly line production 79–80, 84
Aswan High Dam 40
attitudes, effect on technological development 231–2, 242–3
Austin 80
Aztecs 75

Bahmeuller, Charles 56
BAIF Development Research Foundation 55
Balaka lime project 155–6
Bangladesh
    drying technology 122
    landlessness 28
    transportation 39–40, 250–1
Bangladesh Rural Advancement Committee (BRAC)
    credit operation 195, 224, 255
    education 55, 65
    empowerment 58–9
    poultry project 113, 206
    silk project 208–9
Barnett, Andrew 153
Baylis, Trevor 91
Benin bronzes 75
Bielenberg Press 131–6
bilateral donors 60–2, 234
biogas 95–6, 152–3
biomass 139
boats, technology 250–1
Bolivia, credit 255
Boorstin, Daniel 74
Borden, Gail 121
Borlaug, Norman 247–8
Bosnia, unemployment 102
Botswana
    biogas 95–6
    brick production 160–1
    co-ordinated approach 244–5
    renewable energy 138
    silk projects 207
Botswana Technology Centre 242
Brandt Commission Report 5, 8–12, 190–1
Braun, Ernst 241
Brazil 58, 107
brick production 160–2
Britain
    small-scale enterprises 100

Sri Lanka micro-hydro 150, 151–2
  tied aid 237
Brown, David 53
Brundtland Report 6, 12, 37, 191
building materials 154–70
bureaucracy 223, 232
BYS 148–9

Calomiris, Charles 229
calorie intake 24
Canadian International Development Agency (CIDA) 41–2, 189, 237–8
capital 217–18
car production 79–80
carbon dioxide emissions 107, 111
CARE 49, 110–11, 203–5
Carr, Marilyn 206, 211, 225
cement production 156–60, 241–2
Chambers, Robert 23, 25, 196, 257, 261
champions 84, 177, 185–6
charity, history of 50–2
Chen, Marty 206
child mortality 24, 29–30
children, poverty 29–30
China
  cement production 157, 159, 241–2
  foreign direct investment 17
  hydro-electricity 148
  poverty 28
  Three Gorges Dam 41–3
choice of technology 11, 231–8
civil society 56–7
clocks 73–4
Club of Rome 5, 19–20, 190, 191
clustering 100–1, 185
co-ordination 244–5
Coca-Cola 231, 243
coherence 199
Cold War, end of 15–16
Colt, Samuel 83
communications technology 91, 235–6
community development 43
conflicts 15–16
construction materials 154–70
Consultative Group for Internation Agricultural Research (CGIAR) 233
corn milling machines 180
corporate welfare 228–9
corruption 227, 232–3, 252
Costa Rica 240
credit
  informal sector 217–18
  micro-credit 194–5, 203–5, 223–4
  and technology 221–2

Cullis, Adrian 116–17, 118
cultural taboos, women 209–11
customer focus 99

dairying, India 206–7
Daly, Herman 46, 47, 199
dam projects 40–3
DANIDA 234
de Soto, Hernando 223, 224, 232
de Tocqueville, Alexis 51
debt crisis 13, 33
defence spending 15, 261
deforestation 106–7
deindustrialization 74–7
democracy 252–3
Department for International Development (DFID) 151–2, 234
deprivation 25
Desai, Shri Manibhai 55
Desai, Veljibhai 129–31, 177
Development Assistance Committee (DAC) 21, 33
diffusion, of technology 95–6
distortions
  institutional 233–8
  socioeconomic 231–3
Dolorier, Julian 123–4
donor agencies
  arrogance of 193–4
  budgets of 248–9
  choice of technology 234–8
  mistakes of 120, 151–2
  policies of 189, 253–4
dried food 121–4
Drucker, Peter 33
dynamic sustainability 192

Eckaus, Richard 97
economic growth 35
economics of scale 99
education 218–21
  Bangladesh 55, 65
  gender inequalities 202
  and industry 81–2
Egypt 3, 40
electricity, generation 145–53
employment 214–25
empowerment 58–9
  women 211–13
energy, alternative sources 137–9
entrepreneurs 167, 170, 178–81, 217, 220–1
environmental degradation 30–1
environmental movement 190, 255
equity 45, 46

283

exponential growth 46–7
export crops 108–9

factories, assembly lines 79–80
failure, of projects 45, 138, 152–3, 251–2
Fernando, Priyanthi 201
fibre concrete roofing (FCR) 162–70, 241, 250
financial sustainability 189, 193–9
fisheries 76–7, 115
flexible specialization 185
food
    preservation 121–4
    production 20, 108
Ford 79–80, 84
foreign direct investment (FDI) 17
forests
    agroforestry 109–13
    deforestation 106–7
formal sector 101–2, 184
Foster Parents Plan 49
fuel-efficient stoves 139–45
fuelwood 72, 140
Fuller, Buckminster 20
Fulton, Robert 83
function-based strategy 203–5

Gandhi, Mahatma 87–8
Garg, M.K. 157
GATT 10, 14
gender inequalities 200–2
Ghana
    attitudes to local production 231, 243
    Canadian aid 237–8
    dam projects 40–1
    industrialization 10–11
    light engineering 177–86
    metalwork 181–2, 250
    palm oil production 128
    roads 39
    small entrepreneurs 243–4
    vocational training 220
global warming 107
globalization 226–30
GNP per capita 24
governments, and technology 81, 252–3
Grameen Bank 55, 91, 194–5, 203, 224
GRATIS 182–6, 189, 220, 222, 237–8, 250
Greeley, Martin 122
green revolution 6, 10, 104–6, 247–8
groundnut oil 127–8
growth
    limits to 46–7
    strategies 35–7
Guatemala 111

gunpowder 71
Gutenberg, Johann 83

Haiti, agroforestry 110–11
Hancock, Graham 249
handpumps 210, 237–8
Handy, Charles 195
Harper, Malcolm 224
Harrison, Paul 216
health issues 140
Hertsgaard, Mark 42–3
high-yielding varieties (HYVs) 104–5
Hill, Polly 23
Hirschman, Albert 171
household surveys 23–4
housing, building materials 154–70
Human Development Index (HDI) 26
Human Development Report 6, 21, 26, 191–2
humanism 50–5
Huq, Mahbub ul 191–2
hydro-electric projects 148–52

IMF, debt relief 33
In Search of Excellence 99
inappropriate policies 227
inappropriate technology 90, 259
Inca 76
Indefund 155
India
    cement production 157–60
    dairying project 206–7
    irrigation 10
    Kerala fisheries 76–7
    Narmada Project 41
    non-governmental organizations 55
    poverty 24
    silk production 208
    Westland helicopter deal 237
indigenous technical knowledge (ITK) 196
individuals, recognition of 256–8
Indonesia
    economic collapse 18–19
    natural resource depreciation 37
industrial clustering 100–1, 185
Industrial Revolution 183
industrialization
    African failures 172–4
    Ghana 10–11
    very late starters 171–2
informal sector 101–2, 175–6, 216–18, 222–5
    regulations 242
information and communication technology (ICT) 3, 235–6
infrastructural development 37–40

Inger, David 138
Innabuyog 54
institutional distortions 233–8, 243–5
institutional reforms 227
Instituto de Nutricion de Centro America y Panama (INCAP) 123
Integrated Development Association (IDEA) 144
integrated rural development 36, 43–5
Integrated Rural Development Programme (IRDP) 150
interchangeable parts 78–9
intercropping 109–10
intermediate technology
 concept of 86–90
 criticisms of 94–8
 development of 90–4
Intermediate Technology Development Group (ITDG) 90, 95, 97–8
 brick production 161–2
 drying technologies 122, 123–4
 fishing boats 77
 fuel-efficient stoves 143
 livestock services 113–14
 micro-hydro schemes 149–51
 rainwater harvesting 116–18
 sugar technology 125–7
 sunflower oil production 129–31
intermediate technology transfer units (ITTU), Ghana 181–5, 189
Intermediate Technology Workshops 163
International Development Research Centre (IDRC) 134–5, 159
International Institute for Tropical Agriculture (IITA) 111, 233
International Rice Research Institute (IRRI) 105, 233
Internet 3
investment, incentives 239–40
iron working 75, 181–2
irrigation 10
Itaipu Dam 41
Italy, Third Italy 100
Ivory Coast, poverty 24

Jade Mountain Access Company 93
Jamaica 14–15
Japan
 business organization 99
 technology 80, 83
 tied aid 237
jiko stove 141
job creation 10, 101–2, 158, 215
Jolly, Richard 214

Kainji Dam 41
Kaplinsky, Raphael 160, 161, 207
Karachi 64–5, 217
Kenya
 agroforestry 110
 ceramic stoves 141
 livestock development 113–14
 rainwater harvesting 114–18
 sugar processing 125–7
 tile making 164, 165–6, 166–7, 241
 vegetable oil production 134
 women's credit 203–5
Kenya Rural Enterprise Programme (KREP) 55
Kerala 76–7
King, Kenneth 220–1
knowledge transfer 235–6
Korten, David 53

land tenure 27–8, 119
late industrial developers 171–2
lathes 179–80
Latin America, poverty 27
laws 241–2
Layton, Edwin 69
Lever Brothers 231, 243
light engineering 171, 175–86, 254
lime production 155–6
Limits to Growth, The 5, 19–20, 190, 191
Lipton, Michael 247
livestock 113–14
logging 111–12
Lorena stove 142–4
Lutheran World Relief 131, 132

McKee, Katherine 202–3, 204, 211
McNamara, Robert 5, 43, 47
McRobie, George 91
Magaba 175–6
Malawi
 lime production 155–6
 metalwork 176–7
Malawi Entrepreneur Development Institute (MEDI) 220
Manley, Michael 14–15
market fundamentalism 229–30
marketing 243
Martin, Samuel 53
mass production 79
Médecins sans Frontières (MSF) 49
metalwork
 Ghana 181–2
 Malawi 176–7
micro-credit 194–5, 203–5, 223–4
micro-enterprises 223–4

micro-hydro plants 148–52
migration, technical experts 84–5
military expenditure 261
mini-cement plants (MCP) 157–60, 241–2
mini-hydro plants 148–9
minimum tillage 111
missing piece approach 203
Mokyr, Joel 73, 242–3
Morse, Samuel F.B. 83–4
multilateral donors 60–2
Mumford, Lewis 73, 238
Mysore Resettlement and Development Agency (MYRADA) 55

Narmada Project 41
National Council on Women and Development, Ghana 128
natural resources, depreciation 37
Nepal, hydro-power 148–9
Netherlands 111
newly industrialized countries (NICs) 171
Nigeria 41, 173
Nissan 80
Noble, David 79, 82
non-governmental organizations (NGOs) 48–66, 244, 254–6
  and bilateral/multilateral donors 60–2
  and civil society 56–7
  history of 50–2
  overheads 62–4
  proliferation of 57–9
  Southern 54–5, 58, 64–6, 255–6
  tile making 162–8
  typology 52–5
  weaknesses of 49–50
non-tarrif barriers 10, 242
NORAD 115, 234
Norway, aid 76–7, 115, 234
Nowak, Maria 217

OECD 10, 14, 21, 33
official development assistance (ODA) 16, 21
oil (petroleum) 137, 139
oil (vegetable), extraction technologies 127–36
Orangi Pilot Project 64–5
Organization of Rural Associations for Progress (ORAP) 54
Our Common Future 6, 191
Overseas Development Institute (ODI) 23
Oxfam 49, 116–18

Pacey, Arnold 76
Pakistan
  biogas 95

conservation 255
dried apricots 122–3
education 219
high-yielding varieties 105
informal sector 217
non-governmental organizations (NGOs) 55, 64–5
palm oil 128
Palme Commission 6
Parry, John 163–8, 177, 250
participation 98, 196, 257
patents 242
Pearson Commission Report 5, 6–8, 21, 35
Peru 123–4, 223
Philippines, taxation 240
photovoltaic energy 146
policy options 238–9
pollution 20
population, growth rates 19, 30
post-harvest technology 121–36
Postman, Neil 236
poultry, Bangladesh 206
poverty
  alleviation 36–7
  measurements of 22–6
  predictions 32–4, 259–60
  rural areas 26–31
  urban areas 31–2
  war against 246–7
Powell, John 177, 178, 180–1, 182, 185–6
pozzolana cement 159
predictions 258–9
primary commodities 8, 13, 108–9
primary education 218–19
primary health-care (PHC) systems 198
private capital flows 17–19
Probe International 42
PRODEM 255
product refinement 123
Proshika 55
pumps 210, 237–8

rainwater harvesting 114–18
regulations 241–2
Reich, Robert 221
renewable energy 138, 146
Reno, William 222
rent-seeking activity 227, 252
research and development (R&D) 233
rice, high-yielding varieties 105
Riddell, Roger 194, 198, 236
risk avoidance 232
Rondinelli, Dennis 156
roof tiles 162–70, 241, 250

rural development 43–5, 119–20
Rural Development; Putting the Last First 257
rural electrification 145–53
Rural Industries Innovation Centre, Botswana 95–6
rural unemployment 215

Saboo, D.P. 157
Sachs, Jeffrey 238–9
Sarvodaya 55, 142, 210
Save the Children Alliance 49
Schumacher, E.F. 5, 11, 34, 82, 86–90, 92, 96–7, 102–3, 138, 249, 261
Schumpeter, Joseph 99
sectoral approach 206–7, 253–4
Self-Employed Women's Association (SEWA) 55, 203
Shelter Works 166–7
Sierra Leone 16–17, 222–3
silk projects 207–9
SIS Engineering 179–81, 192–3, 223
Six-S movement 54
Small Enterprise Development Corporation (SEDCO) 176
small hydro plants 148
Small is Beautiful 5, 86, 98
Smith, Adam 97
soap production, Ghana 231
Socially Appropriate Technology International Information Services (SATIS) 93
Soko, Frank 176–7
solar power 146
Somalia 237
Soros, George 229–30
South Africa, communications technology 235–6
Sri Lanka
  ceramic stoves 142–5
  Mahaveli scheme 41
  micro-hydro schemes 149–52
  tractors 44–5
static sustainability 192
Steinbeck, John 258–9
Stevens, Siaka 222–3
Stewart, Frances 239–40
stoves, fuel-efficient 139–45
structural adjustment 8, 13–15, 226–8, 260
Suame Magazine 178, 181–2, 185, 219–20
success stories 141–5, 249–50
Sudan, brick production 162
sugar, production technology 124–7
sunflower oil 129–36
sustainability
  agriculture 107–9, 117–20
  concepts of 190–2, 260
  financial 189, 193–5
  human development 192–3
  institutional capacity 197–8
  policy environment 198–9
  stakeholder involvement 195–6
  time-frames 196–7

Tanzania 131–3, 237, 243
tarrifs 10, 14
taxation, incentives 239–40
tea, micro-hydro 149–52
technical education 220
technique, importance 250
technology
  assessment 240–1
  attitudes 231–2, 242–3
  history of 69–77, 242–3
  individuals 83–5
  institutions 80–3, 243–5
  intermediate 86–98
  organization of 78–80
  post-harvest 121–36
  regulations 241–2
  transfer of 69–70, 85
Technology Consultancy Centre (TCC), Ghana 128, 177–82, 220, 222, 250
Tendler, Judith 202–3
Thailand, economic collapse 18
Thant, U 20
Third Italy 100
Thoreau, Henry David 94
Three Gorges Dam 41–3
tied aid 236–7
tiles, fibre concrete roofing 162–70, 241, 250
tinsmiths 176
Tinytech 129–31
tractors 44–5, 240
trade, and technology 70
transfer of technology *see* technology
transportation
  Bangladesh 39–40, 250–1
  and women 201
trickle-down approach 35–6
Turkana rainwater harvesting project 114–18
typewriters 74

ultra-poor 25
UN Conference on the Human Environment, Stockholm (1972) 190
UNCTAD 7–8, 9, 14
underemployment 214–15
UNDP 3, 6, 21, 26, 192
unemployment 101–2, 214–15

287

UNICEF 6, 14, 29, 60
UNIFEM 244
United Kingdom *see* Britain
United Missions to Nepal (UMN) 148–9
United States
   corporate welfare 228
   non-profit organizations 51
   small-scale firms 99–100
   soil erosion 37
urban areas
   poverty 31–2
   unemployment 215–16
USAID 13, 138, 182

Veblen, Thorstein 82
veterinary services 114
vocational training 219–20
Volunteers in Technical Assistance (VITA) 93

war, and technology 70–2
Ward, Barbara 190
wealth creation 36
windmills 147–8
Wokabout Somil 112–13
women
   cultural taboos 209–11
   empowerment 211–13
   inequalities 200–2
   micro-credit 194–5, 203–5
   poverty 28–9

sectoral approach 206–7
silk production 207–9
and technology 200–13
unemployment 215
Working Women's Forum (WWF) 203
World Bank
   appropriate technology 234–5
   debt relief 33
   and NGOs 60–1
   rural development projects 44
   sustainability 198
World Commission on Economy and
   Development (WCED) 191
World Development Indicators 22
World Development Report 6, 12, 22, 25, 32
World Resources Institute 6, 37, 45
World Trade Organization 14
World Vision 49
Worldwatch Institute 6, 46–7

zero tillage 111
Zimbabwe
   brickmaking 161–2
   light engineering 175–6
   regulations 242
   sunflower oil production 128–31
   tinsmiths 243
   unemployment 101–2
Zimbabwe Iron and Steel Corporation 173
ZOPP Ltd 133